Contextual Inquiry for
MEDICAL DEVICE DESIGN

Contextual Inquiry for MEDICAL DEVICE DESIGN

MARY BETH PRIVITERA

University of Cincinnati and Know Why Design, LLC
Cincinnati, Ohio, USA

AMSTERDAM • BOSTON • HEIDELBERG • LONDON
NEW YORK • OXFORD • PARIS • SAN DIEGO
SAN FRANCISCO • SINGAPORE • SYDNEY • TOKYO
Academic Press is an imprint of Elsevier

Academic Press is an imprint of Elsevier
The Boulevard, Langford Lane, Kidlington, Oxford OX5 1GB
225 Wyman Street, Waltham MA 02451

First edition 2015

ISBN: 978-0-12-801852-1

Library of Congress Cataloging-in-Publication Data
A catalog record for this book is available from the Library of Congress

British Library Cataloguing-in-Publication Data
A catalogue record for this book is available from the British Library

For information on all publications visit our website at http://store.elsevier.com

Publisher: Joe Hayton
Acquisition Editor: Fiona Geraghty
Editorial Project Manager: Natasha Welford
Production Project Manager: Melissa Read
Designer: Greg Harris

Printed and bound in the United States of America

CONTENTS

LIST OF CONTRIBUTORS

Tor Alden
HS Design, Inc., Gladstone, New Jersey, USA

Sean Hägen
BlackHägen Design, Dunedin, Florida, USA

Beth Loring
Radius Product Development, Radius, Clinton, Massachusetts, USA

Mary Beth Privitera
University of Cincinnati and Know Why Design, LLC, Cincinnati, Ohio, USA

Andrew Ringer
University of Cincinnati and Mayfield Clinic, Cincinnati, Ohio, USA

Jim Rudolph
Farm Design Inc., Massachusetts, USA

Stephen B. Wilcox
Design Science, Inc., Philadelphia, Pennsylvania, USA

FOREWORD BY THOMAS FOGARTY

The USA has long been the leaders in the development of new technology in the field of medicine, these reasons are multifactorial. This leadership is now under challenge. This book gives insight into basic concepts of how we can maintain our leadership role. The authors give suggestions and pathways that will serve the readers well in maintaining interest and leadership in medical technology. The book covers the subject well. It is written with insight and experience.

Dr. Tom Fogarty

Chairman, Director, Founder of The Fogarty institute for Innovation
National Inventors Hall of Fame for invention of the balloon embolectomy catheter
Presidential National Medal of Technology and Innovation 2014
Mountain View, CA, USA

FOREWORD BY PETER CURRY

Bard Medical Division (BMD) of CR Bard Inc. is a developer and producer of wide ranging critical care products, selling in excess of $700M, globally. Our product development process has embraced the use of contextual inquiry as an essential component to understanding value in the eyes of our customers. It has impacted the development and design of devices used in the treatment of urinary diseases and devices used to precisely control and manage patient temperature after critical life-threatening injury and illness. Our work with Mary Beth and her team has had a significant impact on our development process and directly improved customer satisfaction. We have recently redesigned and launched a product using the contextual inquiry process and a typical response from new and existing customers is "this is exactly what we've been asking for" or "this solution is very intuitive"; which is exactly what we are trying to achieve – organizing extensive customer inputs, derived from intense study of the 'use-environment', into valuable innovation. The products developed using contextual inquiry are having immediate impact on clinicians and patients as well as others in the hospital that make decisions on what to use. In these examples, this commitment to our patients and value to our customer would not have been possible without taking a methodical approach in studying clinical practice as presented in this book.

BMD has worked collaboratively with Mary Beth over the past few years. Her ability to combine objectivity and creativity is a fresh approach to innovation and has improved the early design end of our pipeline. The chapters of this book represent the process she has developed over the course of her career, working collaboratively with the medical device industry and training young professionals. It highlights the relationship between developing a clear understanding of user needs and developing information that informs device design. It is focused on industry practice and incorporates agency requirements.

As engineers, we quickly gravitate to looking for solutions first. Applying the principles highlighted in the chapters that follow, you will understand the importance of, first, answering the "WHY" questions before developing "HOW/WHAT" solutions that often get confused with as starting point of design. Following contextual inquiry and listening to its outputs has shown success in product design for us.

Peter Curry
President, Bard Medical, Inc.
A division of CR Bard
Atlanta, GA, USA

FOREWORD BY WILLIAM S. BALL

Like industry, innovative approaches to medical product design are growing within our universities and colleges daily. The environment is perfect as it combines the efforts of many fields by talented faculty and enthusiastic students who are only bound by the limits of their creativity and ability to innovate. However, to find the right mixture for success requires a serious effort to identify a methodical approach that can turn "imagination into magic." Mary Beth's research in the area of contextual inquiry and the fuzzy front end of design provides us an excellent example of how best to promote medical device innovation across our campus. This book represents years of her dedicated practice both within the academy and as a consultant to the medical device industry. I am very proud of her contribution and her inclusion of our university's Academic Health Center as a true collaborative partner bridging the academic environment with the medical device industry.

Mary Beth's career at our university was in fact groundbreaking from the very start. She was recruited from the college of design to join the engineering faculty of the University of Cincinnati's Department of Biomedical Engineering under my leadership as a full faculty in medical device design. Her role was to integrate design into our developing undergraduate curriculum in medical device innovation. In this role she never disappointed and in fact far exceeded expectations. She rapidly recognized the need for a more experiential and productive approach to the craft, and set the bar for other programs nationally to follow. Over the years, I have watched her bring together the community and its needs with physicians and researchers in the College of Medicine to the benefit of her students, the university and industry. This book comes as no surprise. It highlights the practices she has developed and taught to collaborating biomedical engineering, business and industrial design students as well as faculty in many other disciplines. The book includes detailed descriptions of sound research methodology coupled with a fresh and innovative approach to collaboration. Lastly, the case studies presented offer a validation of the methods presented that can have a lasting impact throughout the device development process.

The University of Cincinnati prides itself on collaboration. This book represents all that our institution has to offer: solving real-world problems through world-class research, building on our local resources to connect with a national and international community, while incorporating innovative pedagogy. This book best exemplifies what we refer to as *Cincinnati Smart,* and Mary Beth represents with pride the very best in what we have to offer. Even better, I have no doubt we are just at the beginning of an important career, with the best yet to come!

William S. Ball, MD
Professor of Biomedical Engineering, Radiology and Pediatrics
Vice-President for Research
Interim Vice-President for Health Affairs and
Dean College of Medicine
University of Cincinnati
Cincinnati, Ohio, USA

FOREWORD BY ARTHUR PANCIOLI

This book contains an in-depth explanation of the contextual inquiry practice and highlights the flexibility of the methodology. When we created our "medical device engine" in the Department of Emergency Medicine and brought together clinicians and experts in contextual inquiry wonderful things happened. Our faculty greatly enjoyed the process of explaining their procedural techniques while specifically highlighting areas of challenge. The clinicians often felt that their processes or techniques had room for improvement, however, they did not have the tools to delineate where the improvements could be found. Working with and observing the clinicians in their clinical arena and debriefing with them frequently, the experts in contextual inquiry were able to extract the gems of understanding that led to device improvement. This process often led to multiplicative benefits as one improvement often brought to light other opportunities for progress. The synergy was remarkable and I am grateful to have had the opportunity to observe the power of this team work and practice.

As readers consider engaging in this type of work you may discover that physicians and other clinical personnel have reservations when they are asked to participate in a process that is predicated on the direct observation of their practice. This was an issue that we encountered in virtually every clinical arena that we explored. Navigating this concern requires a clear explanation of the process, the ground rules and the goals. The case studies from the University of Cincinnati demonstrate the effectiveness of our efforts and success within the academy. The design team involved, under Mary Beth's direction, will always be welcome in our clinical environment. The results of these studies, especially the procedure maps that were generated, provide a truly unique opportunity for learning and represent an amazing dissection of these clinical processes. The depth of analysis and the ability to communicate the intricacies of clinical procedures was simply fascinating.

Finally, I am hopeful many more medical device manufacturers, especially those who design for emergency medicine, take up the practice presented in this book as a proven methodology that improves device design.

Arthur Pancioli, MD
Professor and Chairman
Department of Emergency Medicine
University of Cincinnati, College of Medicine,
Cincinnati, OH, USA

PREFACE

Deep understanding can only be gained through direct experience. The apprentice works alongside a master to gain the skills and knowledge to eventually execute a task on their own; they do more than just ask questions, they observe, absorb and internalize until they understand the nuances that separate adequate from exceptional. Contextual Inquiry (CI) takes on this model by embedding the design team within a users environment. This method is commonly used in consumer products and in the design of software interfaces but is more challenging in medical applications because of limits in access and logistics.

Medical device development involves understanding a complex web of interactions, tasks, and users. Product development teams are challenged to obtain a deeper level of understanding that can only be gained by direct exposure to the user and their environment. The contextual inquiry (CI) methodology provides the research tools and framework to enter the clinical environment and execute observations. These techniques are proven methods that provide a positive and tangible impact in the development of medical devices in every step of the development process.

This book describes the processes of conducting CI in a health care environment, it explains when in the design process it is appropriate to employ the various tools, and provides case studies to illustrate potential applications of CI. This research methodology integrates observation and interview techniques commonly used in existing product development and applies them to this complex use environment. The methodology is consistent with the best practice publication, AAMI TIR 51 Contextual Inquiry. Case studies from the medical device industry are provided which will illustrate and explain main concepts and values.

My objective in writing this book is to share knowledge and experience from the field in the hope that others may accept the methods in their practice ultimately improving medical device design globally.

ACKNOWLEDGMENTS

This book is the result of over 2 decades of practice in the medical device industry and 15 years as serving as a biomedical engineering faculty member at the University of Cincinnati, which included a unique experience while working directly in the Department of Medicine. The methods presented have evolved over the years and have been influenced by many individuals along the way.

I personally thank those in the design community who over many meetings and conferences openly shared and collaborated with me on this text. Their case studies highlight the value of the CI process and their objective feedback has helped assure this test provided actionable guidance through the process. Key contributors to this book include Tor Alden, Steve Wilcox, Beth Loring, Jim Rudolph and Sean Hagen.

Likewise the University of Cincinnati Medical Device Innovation and Entrepreneurship Program students and faculty deserve recognition. The students continue to surprise me with their creativity; their energy is contagious and helped motivate me to write this text. The faculty, Bala Haridas and Jeff Johnson, along with amazing support from Linda Moeller, the academic advisor, have provided the foundation for this outstanding program.

The University of Cincinnati, Department of Emergency Medicine, enabled the formation of a professional team dedicated to collaborating more closely and professionally with the medical device industry; this initiative that brought clinicians, design faculty, and industry together is a truly novel means for improving medical device design. This book would not be possible without this significant commitment to advancing care.

Most of the visuals represented in this book have been either designed or influenced by Kyrsten Sanderson. Kyrsten served as a leader for our professional team and I am grateful to have had the opportunity to work with her. Likewise, Cecilia Arredondo has contributed to the practice through suggesting visual representation of emotion and through her research that introduced playful medical device design.

Thanks to Elsevier for seeing value in its publication with excellent support from program editors, Fiona Geraghty and Natasha Welford.

Lastly, and most importantly, I thank my husband and our three exceptional daughters. Their support for me is endless and very much appreciated.

CHAPTER 1

Introduction to Contextual Inquiry

Mary Beth Privitera
University of Cincinnati and Know Why Design, LLC, Cincinnati, Ohio, USA

Contents

1.1 BACKGROUND AND INTRODUCTION

Contextual inquiry (CI) is a systematic study of people, tasks, procedures, and environments in their work places. It is a commonly used method in user-centered design as the basis for product design decisions and strategy. The method was originally described in the work of Beyer and Holzblatt (1999). The CI research approach is seemingly casual to the study participants, but when done properly, it involves rigorous data analysis and robust determination of the social and physical environments of the workplace through the use of tools that dissect individual task and user behaviors. For the purposes of medical device development (MDD), CI is used to create a body of information about the habits, devices, constraints, and/or systems in the delivery of care. The accumulated body of information can be used to determine product strategies and to assess how current devices are actually used in the field highlighting specific use patterns and behaviors. Additionally, it can be used for the purposes of optimizing existing devices, training, and a means of collaborating with users. The majority of literature

M.B. Privitera: Contextual Inquiry for Medical Device Design.
DOI: http://dx.doi.org/10.1016/B978-0-12-801852-1.00001-0

available on CI methods focuses on user interaction design intended for the design of software systems.

The work of Beyer and Holzblatt has provided a basic framework for CI, with a specific focus on human–computer interface, and has led to a host of research in specific fields highlighting unique user challenges through the process of observation and interview. This book builds on their initial work and is focused on using CI for the purposes of designing and developing medical devices. It describes the processes of conducting CI in various health care environments, including guidance on how to navigate typical challenges with data collection. It further explains, when in the medical device design process it is appropriate to implement this research in order to both achieve optimal design and address new compliance requirements. Case studies are used to illustrate potential processes and applications of CI highlighting research methodologies that integrate observation and interview techniques commonly used in the healthcare environment; these methodologies are consistent with the Association for the Advancement of Medical Instrumentation Technical Information Report #51 on Contextual Inquiry.

The processes presented represent experiences from the author with additional case study contributions from experienced colleagues with full recognition that the methods presented here may not be exhaustive. The fundamental elements of a CI study are that they are flexible and adaptable, therefore evolving. A CI study is rarely completed exactly the same a second time, as those who would be study participants are likely to change and the agenda of discovery has advanced. The breadth of medical device demands for CI is stark; the device spectrum runs the gamut of simple sutures to complex robots, while study goals extend from seeking radical procedure shifts through confirmation of product graphics as a risk mitigation. Additionally, as a general rule variability in CI studies can be high; this is especially true in the practice of medicine as each patient can represent a separate set of challenges. Comparisons can be challenging. The methods presented here are a combination and integration of qualitative research techniques, contextual design, and creativity for the purposes of product design.

1.1.1 CI or Ethnography?

CI and ethnography are terms often used as synonyms. Schuler presents the commonality with ethnography in that "CI is an adaptation of ethnographic research methods to fit the time and resource constraints of engineering" (Schuler and Namioka, 1993). CI is rooted in the importance of analyzing interactions among devices, people, and their workplace. This technique is also a real-time observation of interactions in the real work environment and captures design-informing behaviors and intentions that are not easily gleamed from interviews, surveys, or reports of adverse events alone (Spinuzzi, 2000, 2005).

The design industry started to hire ethnographers in the 1970s (Koshinen et al., 2011). For the purposes of design, ethnography is used to understand the "micro-cultures" or user behaviors relevant to a specific product design within their use context. In the 1990s, when computers moved into the workplace, design began to adopt ethnographic methods for conducting user research (Crabtree et al., 2012). As these methods developed, design teams gained greater appreciation of the real work extended by users as well as the real-time character of each work step within an organization of activities (Crabtree, 1998). Major companies hired several anthropologists in the 1990s, including Apple (1994) and Intel (1996). Design firms, such as IDEO and Fitch, also began to adopt ethnographic field research methods for their clients (Koshinen et al., 2011). With the continuous advent of emerging technologies in the medical field, user research has similarly become more crucial for developing medical products. MDD ethnographic research or CI entails studying the behaviors of specific medical device users in their work environments. Ethnography informs designers who might otherwise design based on their own opinions and assumptions, which they impose their worldview of users (Spinuzzi, 2000).

The practice of ethnography for design requires a technique where the user is observed and interviewed in order to inform design. It is a form of applied social science that draws from sociology, anthropology, and ethnomethodology (Steen, 2011) and leverages an immersive approach that requires analytic assessment. It is not just about going out and observing; an ethnographic research team must have data collection methods, clear objectives, and analytical perspective to assure a methodical process that yields actionable data. Ethnography goes beyond the method of task analysis, wherein the specific tasks and tool uses are discovered. This includes the uncovering of social and environmental factors that affect how someone does their work. In the practice of ethnography, social scientists may dwell for long periods with their study participants. Their immersive data collection period may take years. In contrast, a

CI study is a more focused study for the purposes of product development albeit software or hardware. It uncovers practical action and practical reasoning, and provides empirical topics that can be analyzed while seeking to answer "What do users do and what kind of reasoning is involved in doing it?"(Crabtree et al., 2012). CI is typically a one-way communication from the user to the researcher. In this method, researchers and designers observe users and their actions embedded in social and cultural contexts, while seeking their opinions. The study participant is the mentor, and the researcher, the mentee. The practice of CI is intended to not only inform design but also provide a richness of information to assist in the negotiation of design elements. Of note, there are limitations to this technique as it only describes current practices and may not capture future dreams and perceptions unless specifically queried. For true ethnographic researchers, CI is often considered a less rigorous approach as it is often very qualitative and relies on the researcher to make inferences about the situation.

CI is a form of ethnography. Ethnography is a broader methodology that encompasses more investigative techniques than interviewing and studying user in their work environment. Figure 1.1 illustrates that both research techniques use the same methods and build models; however, their purposes are different. Ethnography has a more broad approach, whereas CI is focused on informing design. CI is like ethnography in that

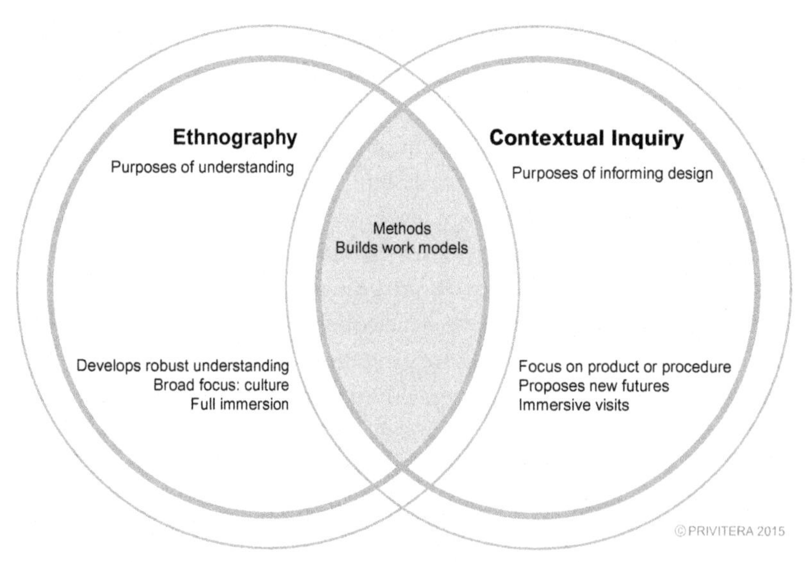

Figure 1.1 Differences between ethnography and contextual inquiry.

the methods include observations and interviews of a particular work group and break down their steps, behaviors, and opinions in a methodical manner, then develops various models to develop an understanding of the user. However, unlike ethnography, once those work models are understood and diagrammed, they are redesigned with a new perspective (Spinuzzi, 2000, 2005). Thus envisioning a new future. In MDD, this new perspective is product opportunity and definition.

CI methods involve part observation and part interview in the users' workplace, while they complete a task: note the similarity to the description of ethnography. It goes beyond observation to making inquiries regarding specific tools or procedures, while the user is interacting in their workplace. Spinuzzi deduces CI is explicitly structured as a field method oriented toward design. It involves short, targeted observations, and conversational interviews coupled with elaborate analyses guided by work structure models (Spinuzzi, 2000).

CI has the purpose to uncover explicit and observable knowledge about context of the product use and user. Mattelmaki et al. (2005) provide a definition of CI, in which there is a focus on the definition of what kind of information being sought (context) and how that information is elicited (using generative techniques), that is, in turn designed information for the purposes of use by design teams (creating a context map) (Mattelmaki et al., 2011). Studying context helps designers gain empathy for users to avoid fixating on preset assumptions about the user and the product, and to create concepts on how a product can be experienced.

The process of conducting CI includes scope definition, literature review, planning, data collection, build analysis tool, data entry, compile evidence, then visualize (Figure 1.2). The process of analysis includes the researcher study the data, search and look to be surprised, find patterns, and create an overall view.

The number of participants and the depth to which a subject is studied determines intensity of the inquiry (Mattelmäki, 2005). The process and fidelity of conducting CI is dependent on the careful selection of participants, the study design itself, and the diligence of the researcher. Interviews are ideally conducted in the user's actual workplace wherein the researcher watches users do their own work tasks and discuss any artifacts they generate or use with them. In addition, the researcher gathers detailed stories of specific past events when they are relevant to the project focus (Holtzblatt et al., 2005a). A goal of this type of research is to partner the user and the researcher in order to understand the user's

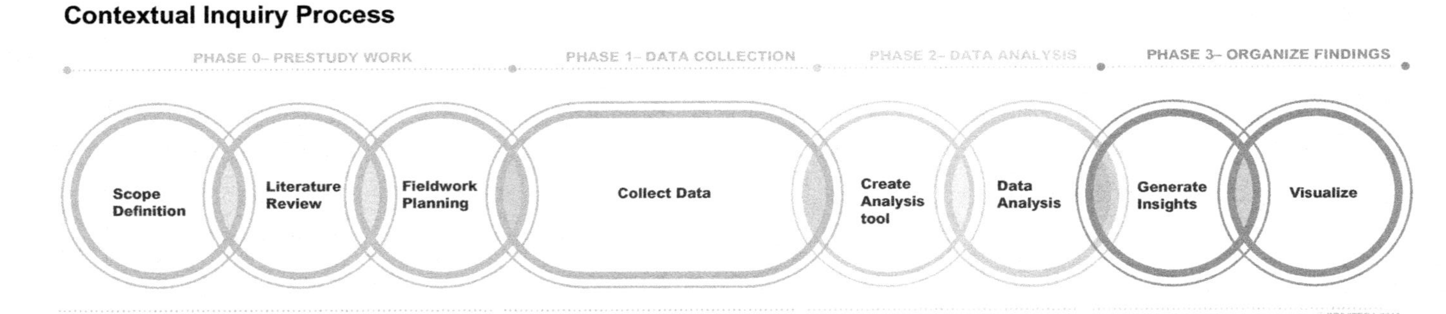

Figure 1.2 Overview of the contextual inquiry process.

work. The interview alternates between observing the user as they work and discussing what the user did and why. Finally all data are subject to interpretation and analysis. In some instances, the researcher may share their interpretations and insights with the user during or subsequent to the interview helping expand or correct the researcher's understanding (Holtzblatt et al., 2005b).

There are two roles a researcher can adopt in CI techniques: observer-as-participant and participant-as-observer. In the instance of the observer-as-participant, most of the work is observation, but the researcher interacts from time to time with those being observed, perhaps to clarify something. In the instance of the participant-as-observer, the researcher participates in the day-to-day life of the situation being studied, but is known that she or he is carrying out research, and there is an element of observing the behavior without participating in it (Bechhofer and Paterson, 2012).

In summary, CI techniques are a proven methodology that inform the design team and promotes user involvement. Through this methodology the design team can get first-hand knowledge of the users' environment, the personalities of all stakeholders involved, and produce initial requirements or recommendations for design. They provide the theoretical framework for observing people in their actual environment, while CI provides a methodology specifically targeted toward the design and engineering disciplines that involve principles of ethnography. The two are closely related. This book is focused solely on the practice of CI for MDD.

1.1.2 Purpose and Rationale

For the purposes of product design, CI provides a deeper understanding of users, use environments, tasks, and procedures than what can be achieved through interview-based methods alone. CI provides information about the constraints that a new device or system must operate within while providing additional information concerning unmet user needs. In addition, the process can uncover and identify problems with existing devices and systems that a new device can address. As a result, this information can readily create a robust task analysis.

The viewpoint of those who design and manufacture devices is often different than those who use devices. Developers build their understanding based on ideal circumstances, design intent, generally accepted practices, and filtered feedback. Users build their perspective relative to their personal experiences, which may or may not reflect general practices, and which are specific to their individual background. CI methodology gathers observation and interview data, facilitating objective analysis of the behavior and opinion of the user. This process will enable device developers to have empathy for users and their needs. This in turn will assist in providing evidence for design decision-making.

1.1.3 Background of MDD

A medical device is a tool that may be used to diagnose or treat a disease, condition, or assistive need on or by humans. In both the USA and European Union (EU), the definition of a device is broad, and ultimately, it is the responsibility of the manufacturer to work with the regulating entities to determine appropriate registration and approval.

The US Federal Drug Administration defines a medical device as:

"... an instrument, apparatus, implement, machine, contrivance, implant, in vitro reagent, or other similar or related article, including component part, or accessory which is: recognized by the national formulary, intended for use in the diagnosis of disease or other conditions, or in the cure, mitigation, treatment, or prevention of disease, in man or other animals, or intended to affect the structure or any function of the body of man or other animals, and which does not achieve any of its primary intended purposes through chemical action within or on the body of man or other animals which is not dependent upon being metabolized for the achievement of any of its primary intended purposes." (US FDA, n.d.)

The European Union defines a medical device as any article/s that are:

> *... intended to be used for a medical purpose are considered a medical device. The manufacturer assigns the medical purpose to a product and thereby declares the product a medical device. The manufacturer determines through the label, the instruction for use and the promotional material related to a given device its specific medical purpose. As the directive given aims essentially at the protection of patients and users, the medical purpose relates in general to finished products regardless of whether they are intended to be use alone or in conjunction with others. Therefore products intended to have a toiletry or cosmetic purpose are not medical devices even though they may be used for prevention of a disease e.g. tooth brushes, contact lenses without corrective function, bleaching test products, and instruments for tattooing. Included in this description is the software that influences the proper functioning of a device. Software related to the functioning of a medical device may be part of a device or a device in its own right if it is placed on the market separately from the related device.*
>
> *(European Commission, 1994)*

The descriptions above demonstrate subtle nuances, such as the definition for intended use, between these primary agencies and describe the broadness in definition of a medical device. One difference between the two regions is inclusion of use on any animal in the US definition (Ogrodnick, 2013). Regardless of the region, agency approval for license to market is often required and comes with the significant requirements, such as proof of efficacy, safety, and usability.

A commonality of all devices is the application of science as a foundation in order to diagnose or mitigate a symptom or disease. The influence of basic science and the product development process are intertwined with one, not being more important than the other, just more or less active across the development and discovery cycle (see Figure 1.3). Figure 1.3 highlights that basic science forms the fundamental theory that explains our lives and overtime design, and engineering applies science to improve our lives through the commercialization of devices.

Due to this, the MDD process requires a close relationship between the clinical practitioners, the scientists, and the development team (Figure 1.4). The current design process is a combination of methods from engineering disciplines, government regulatory agencies (domestic and international) and independent certification, and compliance companies. The goal of the processes is to be certain that a new product meets the expectation of the user, is safe and effective in providing its claimed benefits (Gilman et al., 2009).

Figure 1.4 illustrates the cycle from a new technology/new problem exploration through to clinical trial and the efforts required to commercialize

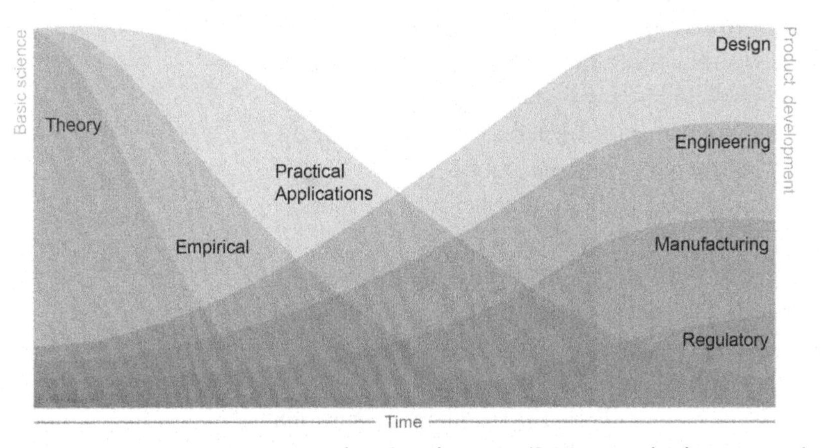

Figure 1.3 Basic Science versus Product Development (Privitera and Johnson, 2009).

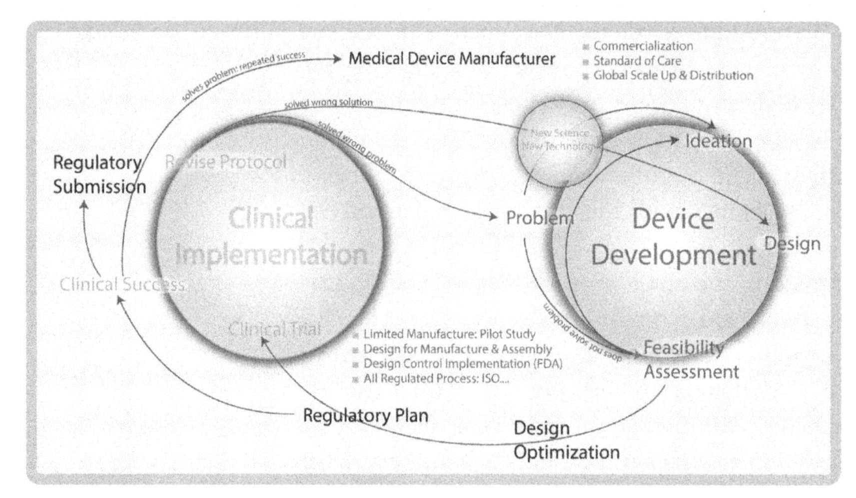

Figure 1.4 Relationship of the clinical environment to device development.

a medical device. This diagram is the result of internal gap analysis within the University of Cincinnati by the Center for Clinical, Translational, Science and Training. In Figure 1.4, the process is highly cyclical with two main emphasis: device development and clinical implementation. For a new device to be used within the clinical environment, certain hurdles must be achieved such as a regulatory plan and design optimization. With any new technology, the requirement of solving a valuable problem, the freedom of trial and error in the safety of a laboratory, coupled with a follow-up exploration as to why something worked or did not work, is essential in MDD. Ultimately, the success of a device not only relies on the strong application of science (both medical and engineering) but also partnerships between those involved in the practice of medicine and the device-manufacturing entity. There is very little activity within the cycle that is completed by any one individual entity and/or without consideration from either the regulatory agencies, the manufacturing, or the clinical entity. In essence, collaboration is essential for MDD.

A typical device development process is the traditional waterfall model (Figure 1.5).

This very basic model of product development is promoted by regulatory agencies and is widely used. The development proceeds in a logical sequence of phases or stages. The phases include user needs, design input, design process, design output, and medical device release. A review is conducted at each phase of development. During design input through the design output phases, the design is assessed to assure that the design

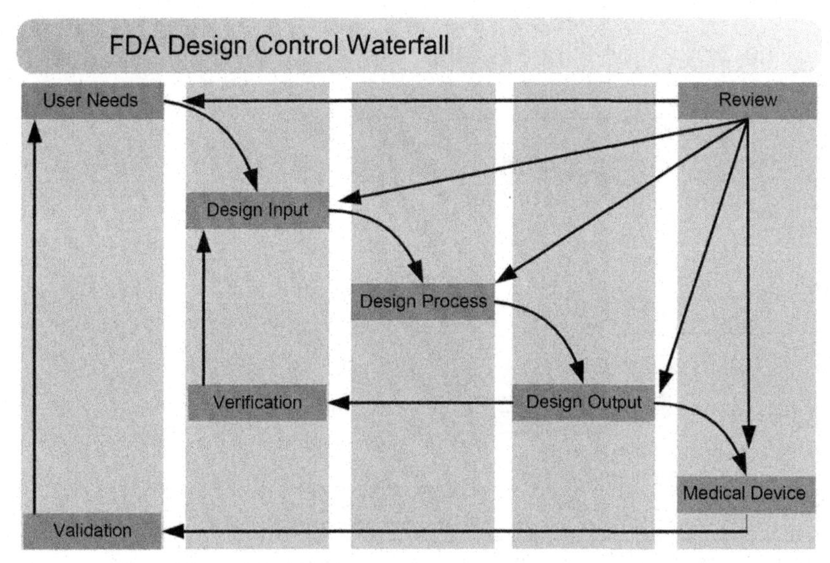

Figure 1.5 Waterfall model of MDD.

meets the user needs and requirements as determined in the initial phases. This is called design verification. Lastly, once the device has undergone a full development and manufacturing has begun, final assessments are conducted. This is called design validation. Throughout the process, the entire development team is required to maintain the "DHF," which documents and describes in detail about idea, concept, test, and review.

All regulating bodies globally mandate a formal development and manufacturing process, wherein it is possible to trace design research to device criteria and subsequent device design. All steps within the development process must be documented following a quality system regulation to assure the design history not only supports a product release but also provides a solid reference for future changes.

While the FDA provides a basic framework on what must be delivered, developers are left to create their own procedures to define how to meet those deliverables. Design control is the formal development process, the fundamental requirement for regulatory approval. Simply stated, design control requires a documented history of development and the origins of any decision made during the development process. The document trail is required, which takes on a structured approach in order to produce a technical file of sufficient rigor to be accepted in an agency audit (Ogrodnick, 2013).

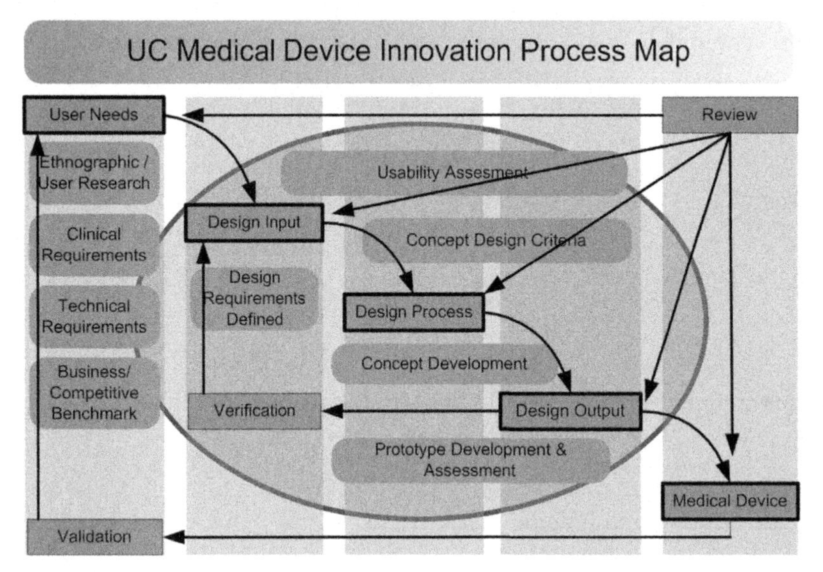

Figure 1.6 US FDA Design Control Process defined by the University of Cincinnati Medical Device Innovation and entrepreneurship Program (Privitera and Grood, 2005; Lewis, 2010).

Figure 1.6 describes the waterfall process promoted by the US FDA in boxes (with heavy outline) with an overlay of a typical design process developed at the University of Cincinnati (Privitera and Grood, 2005). The framework of this model meets US Federal Regulation 820.30 titled "Design Control," which is a formal product development methodology. During an audit by a regulating agency, each section of design process model is assessed for the prior section. The results are recorded in a design history file (DHF) that in turn becomes the main record of "design control."

Design control is simply one of many requirements that a medical device must meet in order to be sold, but ultimately those compliance rules are all intended to facilitate the most basic requirement—that the device satisfies the requirements of the clinical treatment with proven efficacy, while promoting safe use and incorporating usability features in order to meet agency standards. Unfortunately, the myriad of compliance requirements often are seen as an end in themselves as opposed to a tool to achieve the ultimate goal. As a reminder of the ultimate goal, AAMI/ ANSI/IEC Standard 62366, *Medical Devices – Application of Usability Engineering to Medical Devices*, specifically calls for user research as a validation method demonstrating usability of a medical device (International

Electrotechnical Commission, 2007). This international standard promotes the involvement of the use from the idea conception, through the initial stages of development and ultimately a formal assessment of usability demonstrating ergonomic design features. The FDA recognizes CI studies as a means to the identification of user needs and thorough understanding of how users interact with the medical device in the actual use environment. Additionally, the Association for the Advancement of Medical Instrumentation Human Engineering Standard 75 indicates that the product design itself take the environment of use in consideration during development (AAMI, 2009). For example, if a medical device is used in a loud room, the alarms and audible signals that provide feedback to users must be specified accordingly to maximize the usability of the device. This understanding of the environment can only come from actually visiting the environment, attaining user input, and observation prior to finalizing design as well as the conduction of a usability validation study. These efforts develop the human factors dossier, are an integral part of the DHF, and hopefully provide a reminder throughout the development process that all research should be user and patient centric.

1.1.4 When is CI Completed in MDD

CI is most commonly conducted during phase zero prior to the onset of design control (Figure 1.7). Phase zero is an exploratory phase dedicated to conducting research and experimentation for the purposes of concept ideation and assessing the net present value (NPV). NPV is the calculation of the amount of investment required and the expected return. It is a predictive model and the more time spent defining value and refining estimates the better. A CI study conducted in this phase can have a lasting impact throughout the development process.

Note, while this model has clean delineations between phases, in reality, the product development phases are often blurred from one to the next. As such, Figure 1.7 is a simplified model and the activities, as presented and divided, may vary between companies.

Table 1.1 brings together the *potential impact of a CI* study with phases of product development. A CI study builds the foundation of design information from the onset of the development cycle. It has the most impact in the first 3 phases of development and can generate partnerships with users that carry on throughout the entire development cycle with the main goal being to ultimately improve the usability of the device

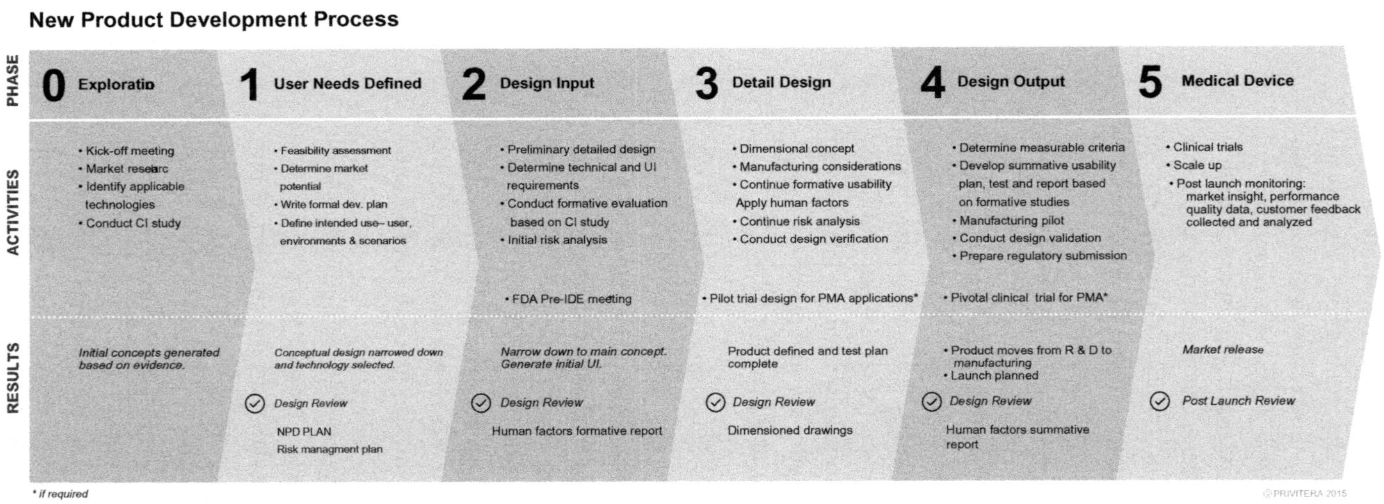

New Product Development Process

PHASE
ACTIVITIES
RESULTS

0 Exploratio
• Kick-off meeting
• Market researc
• Identify applicable technologies
• Conduct CI study
Initial concepts generated based on evidence.

1 User Needs Defined
• Feasibility assessment
• Determine market potential
• Write formal dev. plan
• Define intended use– user, environments & scenarios
Conceptual design narrowed down and technology selected.
Design Review
NPD PLAN
Risk managment plan

2 Design Input
• Preliminary detailed design
• Determine technical and UI requirements
• Conduct formative evaluation based on CI study
• Initial risk analysis
• FDA Pre-IDE meeting
Narrow down to main concept. Generate initial UI.
Design Review
Human factors formative report

3 Detail Design
• Dimensional concept
• Manufacturing considerations
• Continue formative usability Apply human factors
• Continue risk analysis
• Conduct design verification
• Pilot trial design for PMA applications*
Product defined and test plan complete
Design Review
Dimensioned drawings

4 Design Output
• Determine measurable criteria
• Develop summative usability plan, test and report based on formative studies
• Manufacturing pilot
• Conduct design validation
• Prepare regulatory submission
• Pivotal clinical trial for PMA*
• Product moves from R & D to manufacturing
• Launch planned
Design Review
Human factors summative report

5 Medical Device
• Clinical trials
• Scale up
• Post launch monitoring: market insight, performance quality data, customer feedback collected and analyzed
Market release
Post Launch Review

* if required

Figure 1.7 Overview of activities, including predesign control, in the phases of MDD.

Table 1.1 Impact of CI studies defined for each phase of device development

Phase	Title	*CI impact* within each Phase
0	Exploration	*Define new opportunities, define user behaviors, define use environment, define social structure around device use*
1	User Needs	*Develop fundamental design requirements in the words of the user based on their values.* The market potential, use, intended user and use environment are fully described and the technical requirements are included. A need statement is a concise description of a goal (what the device needs to do) but does not indicate how to achieve the goal. It can also include qualitative targets and use descriptions
2	Design Input	*Inform storyboards of device, identify partners for preliminary formative evaluations, and other techniques that further design exploration and definition* During this phase best practices suggest that formal design verification and risk analysis begins with emphasis on preliminary testing of conceptual designs and identification of potential application risks that should be mitigated through design
3	Detail Design	*Inform design regarding desired user interface, inform risk analysis, identify partners for further formative evaluations* as the design is finalized into dimensional concepts Manufacturing considerations and further risk and usability assessments take hold
4	Design Output	*Provide root traceability of human factors* application at the onset of the program while conducting studies focused on assuring the design is robust Preparing for regulatory submission and manufacturing pilots
5	Medical Device	*Improve usability* and reduce likelihood of use errors, reduce issues submitted to post-market surveillance One important point to note is the additional requirement of post market surveillance by regulating agencies in order to provide real-time assessment of the risks and benefits of a medical device. This requirement further emphasizes the need to focus early design efforts with the user in mind

through informed design. For medical devices, informed design means that the design itself is based on solid clinical, technical evidence that is appropriate for the user and use context.

1.1.5 Uses of CI in MDD

The fundamental tenets of medical device design are functional efficacy and safety, and the regulating agencies now expect formalities in design and documentation that assure human factors and applied usability testing are part of the design process, to assure the user and the environment of use are understood. CI facilitates a thorough understanding of every user who comes in contact with a device and the various environments a device may be used. In healthcare, the user and the use locations vary widely and sometimes change throughout device life cycle. For example, a left ventricular assist device , that is, an electromechanical assist device that partially or completely replaces a heart, has a complex set of users beginning with the surgeon who places it, the patient as well as nurses. Secondary users often do not have a role in the selection of device use; however, their ability to identify a device, prepare it for the primary user, and interact with the device is of equal importance to the primary user. For example, a surgical physician's assistant who fumbles during the preparation of a device due to difficult usability may encounter a less than happy surgeon. The surgeon and the patient are both impacted in this situation. The surgeon gets frustrated as the procedure is taking longer and they are waiting. The patient must be under general anesthesia longer and/or other complications may ensue due to response time. The CI process enables the research team to observe and interact with all users and study the social and environmental dynamics that can play a role in new device development. Areas of critical conditions within procedures are examined closely to assure all the needs are identified.

Figure 1.8 shows a typical surgical team, in this case a cardiac surgery team, identifying a priority of device development for each stakeholder.

CI can Determine Usability Objectives through the Study of Behaviors

Inevitably users are not a homogenous group, they vary according to demographic, capabilities, and limitations. CI will assist in developing subsequent usability objectives to be used later in the design process. Ultimately, a study can determine behaviors that are fundamental tenets and should be considered in product design. This includes the identification of all tools used and the opinions of the user. For example,

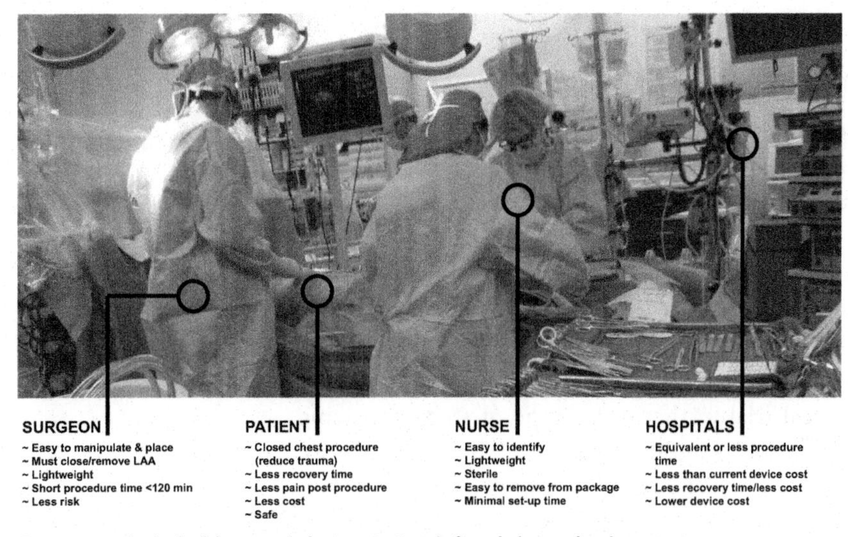

Figure 1.8 Stakeholders and their priority defined through observation.

a fundamental task of an anesthesiologist in complex cases is to prepare medications in various syringes prior to the start of a case. This process is directly in control of the physician and relies on their knowledge as to what medications they prefer, for what reasons, and the doses required based on patient presentation. A potential finding could be that taking this control away could be detrimental; it could be the very activity they enjoy about their job. Additionally, CI can be used to determine use behaviors of competitive products. This can identify the use considerations of ancillary devices or anatomical/physiological phenomenon, that is, assessing the overall procedure, with all devices used not solely the one (we) are responsible for designing. Information gathered could lead to an understanding of why behaviors are exhibited.

CI can Assist in Developing Business Strategy

Determining what should be developed and where is the biggest challenge area, that is, of an opportunity, can be determined through a CI study. In the practice of medicine, challenge does not always equate opportunity. There are instances where the clinical problem is quite difficult and requires elements of bravery and confidence on behalf of the clinical practitioner. Not everything they do is easy, nor will it be. The emotional commitment that enables difficult life-saving measures undertaken may be the very essence of value in the eyes of the practitioner. A CI can assess perceived value through the discovery of culture in the workplace.

Similarities and Differences to a Clinical Trial

Clinical trials are prospective biomedical or behavioral research studies involving human subjects who are designed to answer a specific question about specific interventions, such as vaccines, drugs, treatments, devices, or novel approaches. They are intended to generate safety and efficacy data and require regulatory approval prior to execution. Typically, there is a pilot study that progressively increases in size as data are collected and compared to a specific level of statistical relevance.

A CI is similar to a clinical trial in that it is a type of study involving human subjects. For the purposes of MDD, they are focused on devices, the context of their use, and the social structures found within. There is typically a pilot study, and the data grow significantly with each site visit. In contrast, most clinical trials are designed for publication in a peer-reviewed journal and require ethics approval. For CI studies in medical devices, the intent is to inform design typically with a competitive organization. As such, the results are rarely made public. In fact, for the purposes of this book, the case study on diagnostic angiograms presented in detail originated in the academic environment with approval to publish. The respective device companies have approved the content found in the other studies; however, precise design insights are not included. Thus, the requirement of ethics approval is not clear. Lastly, there is a trend of conducting large multicentered randomized clinical trials with significant statistical analysis. In CI studies, there is very limited statistics that can be applied to the data. The data are qualitative in nature and are used for the purposes of driving design.

1.1.6 Arguments for or Against Conducting CI

"We Know Our Customer. We Do Not a Formal Study to Tell Us What We Already Know"

Most MDD teams have a belief that they are experts in knowing everything there is to know regarding device use on the device they design. Often this belief is derived from prior individual experiences or from a long-standing relationship that the company has with its product lines. As time goes on, these beliefs get more deeply rooted to the point where they no longer are a perspective, but are defined as facts. Coupled to this, is that the origin of the development projects typically do not begin with a solid understand of the "WHY" behind the product (i.e., "accessing the target anatomy is challenging because of orientation of adjacent structures"), rather they begin with the explicit request for "WHAT" will be delivered (i.e., "we need an articulating device") because the sales team or an influential physician has requested it.

Its likely true, many medical device companies have in-house experts who have designed the same device or for the same disease state for many years. They hold strong opinions as to how their customers use their devices. A fact of the matter is that users are always evolving and are creative. Not all challenges are communicated back to suppliers and not all serious issues get reported during post-market surveillance. The users themselves may not communicate behaviors they do everyday as this can make awkward conversation or be too personal in nature. Conducting a CI study with participant de-identification can protect confidentiality issues and uncover those behaviors that are overlooked as routine and insignificant to users but may inform design.

"It Is Too Costly"

The cost of conducting a CI study is dependent on the team size, the number of site visits, required honorariums, and any travel expenses. They can get quite costly if the study goal is to determine global perspective. The costs of conducting the study must be balanced with the benefits of its results. Beginning the design process off with a clear understanding of the user and use context with a shared vision across the entire team will assist in better design decisions, build a comprehensive human factors dossier, and improve overall product design.

"Participants are a Pain"

Access to study participants can be challenging. Additionally, the participants themselves may be shy. They may be uncomfortable being observed and may not understand why someone is there or why what they say is important. Most clinical practitioners are not familiar with MDD processes. This may require repeated explanation and approvals throughout the data collection phase. Lastly, study participants might be concerned with where and how the information and images collected will be used. All of these realities can be balanced with the opportunity that through the CI process additional partnerships with study participants may be gained, which can help throughout the development process or on other development programs. Lastly, participants may have as many problems as they do ideas or subtle device improvements that can be captured in this process. The end deliverable from a CI study is to capture user needs in the form of design insights, design considerations, and design ideas. Ideas may be sparked through the conversations with participants, and the development team will have the opportunity to better understand the value of the ideas in real time and in the view of their customer.

"The Data Collected Is Overwhelming and I Have No Idea What to Do with It"

There will be challenges in the field. These challenges may include cancelations, delay in scheduling, contacting the wrong person, denied access due to lack of preparation (could be the team or the site), or just denied access due to the lack of going through the appropriate channels for access. Additionally, there is a point of collecting too much data, and in data analysis, the team may find themselves swamped with data without gaining many new insights. These issues can be resolved with program management and adaptability by the research team. The research and the design team should work closely together to maximize the overall benefit of the study.

A reality of a CI study is that once you have done a first study these challenges get easier; the team has a system of data collection, management, and analysis. Additionally, the studies provide a chance to develop simulations and continue the relationship with participants. The ability and opportunity to involve users directly has value beyond the specific project need and can influence the overall product development process.

"It Takes Too Long and We Cannot Afford the Time for a Study"

CI studies can come in all lengths and depths depending on the needs, and in some instances the cost of not doing the work can be significant. For example, products have failed due to a lack of understanding of true needs by the user. Or over time devices have become so entrenched that they are difficult behaviors to change and require agency intervention. An example of this is luer connectors. Luer connectors are commonplace in every location within hospital and are used to connect needles to syinges, catheters to IV lines as well as feeding tubes. Unfortunately for patients the same standard connection is used in every instance. This means, it is possible for a clinical provider to connect a feeding tube to an IV line: the result can be death. Currently, these connectors are going through very complex and costly changes because the human factor issues were not addressed at the onset of the development programs.

Conducting a CI study during phase zero is ideal. However, in some instances programs may have already started. Unless the design is frozen with the inability to make changes to it, a CI study can be incorporated into the process. A CI study is intended to be flexible and adaptable. The methodology provides a framework by which a development team can work with users. Due to agency requirements, this is an expectation for inclusion and best practice.

1.1.7 Starting Off with a CI Study Makes Sense

The goal of any product development program is to design a product that is useful and desirable. The same is true for MDD. The CI process is a proven methodology in the user-centered design process developed for consumer products. Additionally, regulating agencies promote user involvement through the development and recognition of industry standards. Conducting a CI study can achieve both: the determination of value while beginning to document user needs according to design control requirement.

REFERENCES

AAMI, 2009. ANSI/AAMI HE75, 2009/(R)2013. Human factors engineering—Design of medical devices, USA.

Bechhofer, F., Paterson, L., 2012. Principles of Research Design in the Social Sciences. Taylor and Francis, Hoboken.

Beyer, H., Holzblatt, K., 1999. Contextual Design Defining Custoner-Centered Systems. Morgan Kaufmann, San Diego, CA.

Crabtree, A., 1998. Ethnography in participatory design. Proceedings of the 1998 Participatory Design Conference, pp. 93–105.

Crabtree, A., Rouncefield, M., Tolmie, P., 2012. Doing Design Ethnography. Springer, New York.

European Commission. 1994. Medical Devices: Guidance document. Available at: <http://ec.europa.eu/health/medical-devices/files/meddev/2_1_4____03-1994_en.pdf>.

Gilman, B.L., Brewer, J.E., Kroll, M.W., 2009. Medical device design process. Conference Proceedings: Annual International Conference of the IEEE Engineering in Medicine and Biology Society. IEEE Engineering in Medicine and Biology Society Conference, 2009, 5609–5612.

Holtzblatt, K., Wendell, J.B., Wood, S., 2005a. Storyboarding. In: Rapid Contextual Design. Interactive Technologies. Morgan Kaufmann, San Francisco, pp. 229–243.

Holtzblatt, K., Wendell, J.B., Wood, S., 2005b. Rapid Contextual Design. Elsevier, San Francisco.

International Electrotechnical Commission, 2007. IEC 62366:2007 – Medical devices–Application of usability engineering to medical devices.

Koshinen, I., Zimmerman, J., Binder, T., Redstrom, J., Wensveen, S., 2011. Design Research Through Practice From the Lab, Field, and Showroom. Elsevier, Waltham, MA.

Lewis, R., 2010. A problem well defined is nearly solved. J Med Devices 4 (2), 027503-027503-1.

Mattelmäki, T., 2005. Applying probes—from inspirational notes to collaborative insights. CoDesign 1 (2), 83—102.

Mattelmaki, T., Brandt, E., Vaajakallio, K., 2011. On designing open- ended interpretations for collaborative design exploration. CoDesign 7 (2), 79—93.

Ogrodnick, P., 2013. Medical Device Design: Innovation from Concept to Market. Elsevier Inc., London.

Privitera, M.B., Grood, E., 2005. Robotic surgery: results of an ethnographic and design research program. NCIIA 9th Annual Conference, NCIIA, 18–20 March 2004, San Jose, CA, USA.

Privitera, M.B., Johnson, J., 2009. Interconnections of basic science research and product development in medical device design. Conf Proc IEEE Eng Med Biol Soc, 2009, pp. 5595—5598.

Schuler, D., Namioka, A., 1993. Participatory Design: Principles and Practices. L. Erlbaum Associates, Hillsdale, NJ.

Spinuzzi, C., 2000. Investigating the technology-work relationship: a critical comparison of three qualitative field methods. In 18th Annual Conference on Computer Documentation. Technology and Teamwork. Proceedings IEEE Professional Communication Society International Professional Communication Conference. pp. 419—432.

Spinuzzi, C., 2005. The methodology of participatory design. Tech. Comm. 52, 163—174.

Steen, M., 2011. Tensions in human-centred design. CoDesign 7 (1), 45—60.

US FDA, n.d. Classify Your Medical Device — Is The Product A Medical Device? Available at: <http://www.fda.gov/medicaldevices/deviceregulationandguidance/overview/classifyyourdevice/ucm051512.htm> (accessed 18.01.15.).

CHAPTER 2

Planning a CI Study for Medical Device Development

Mary Beth Privitera
University of Cincinnati and Know Why Design, LLC, Cincinnati, Ohio, USA

Contents

M.B. Privitera: Contextual Inquiry for Medical Device Design.
DOI: http://dx.doi.org/10.1016/B978-0-12-801852-1.00002-2

2.1 OVERVIEW OF CI STUDY

Contextual inquiry (CI) studies in medical device development (MDD) focus on the "what" and the "why" while in the field, and then through analysis and interpretation the "so what?" that drives medical device design can be determined. The process is based on a set of principles that allow it to be adapted to each situation studied (Beyer and Holtzblatt, 1998). This is achieved through visiting with medical device users as they use the devicesand developing a rapport for the sharing of methods and opinions. The data are collected and interpreted for multiple purposes, including opportunity identification. Conducting a CI study in this environment requires formal planning. However, the actual conduction of the study and interaction with the participants is largely informal. Like best practices in MDD, the best results happen when a multidisciplinary research and design team are involved at every step. Each discipline brings expertise and lens through which they view the clinical problem and will add value to the research.

CI is a flexible and adaptable method; however, there are common steps for conducting a study in support of MDD. These include (Figure 2.1)

1. Scope definition
2. Literature review
3. Fieldwork planning
4. Data collection: pilot
5. Prepare data analysis tools
6. Data Analysis
7. Compile evidence and generate insights
8. Visualize data

This chapter describes an overview of each step in a CI study for MDD. It discusses the importance of secondary research, determining appropriate sites and participants. Study preparation by the research team includes the necessary training and certifications as well as Institutional Board Review requirements. Study protocol development is discussed in detail with multiple examples. Finally, a summary on conducting CI studies outside the United States is discussed.

2.2 PREPARATION AND SETTING THE BACKGROUND

In conducting a CI study, there will always be a degree of uncertainty that is welcomed but can be stressful to a research team. The uncertainty

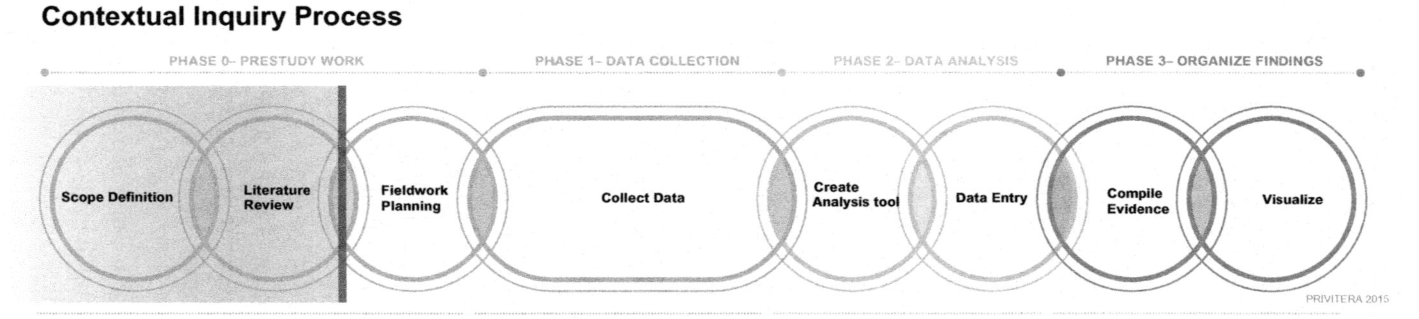

Figure 2.1 A CI study in medical devices requires preparation and background study.

is precisely what a CI studies: it is watching users as they perform their tasks without unduly influencing the outcome. In watching a clinical procedure in order to observe a medical device being used, it is easy to get lost, confused, and have no idea what is really going on unless the researcher has spent time dedicated to gaining the knowledge necessary to begin asking the questions. The equivalent of this is poor medical-based television programs wherein the use of devices, terminology, and clinical procedures is not accurate to real practice, yet the audience has no idea. The research team is the audience and they have a responsibility for the success of the study to gain knowledge in anatomy, typical patient presentations or diagnostic conditions as well as any procedure protocols that are readily available in published literature. The uncertainty in a healthcare CI study should be the interactions and behaviors of the user and not the procedure itself.

2.2.1 Beginning where Others Left Off

Device companies, especially large companies, typically have a wealth of information regarding their customers, their devices, and their uses within the organization. Spending time conducting research that has already been done is a waste of time and valuable resources. Before launching into a new research program, it is imperative to understand what work has already been conductedand the results and potential impact it may have for a CI study. For example, many marketing departments will have data regarding their customers. They have an understanding of the market; however, they lack the understanding of what exactly to design (Beyer and Holtzblatt, 1998). Marketing information and design information are different pieces of information. Each is valuable and often interrelated. Almost always there is marketing data available from within an organization that can assist in jump-starting the secondary research. However, there is no harm in gathering further information prior to the onset of a CI study.

As mentioned, designing a medical device requires clinical knowledge. The ability to tap the mind of the internal scientist or technologists who have tenure at the company can advance the scientific knowledge of the research team. Many times these internal resources are aware of published studies or projects the research team would struggle to find. Starting from what is known, confirming the knowledge and building upon the foundation should be the goal of a CI program. Typically this wealth of knowledge can be used to determine key opinion leaders (KOLs) who may be target study participants, other sites of interest, provide the research team with anatomy and procedural knowledge, and identify gaps in the knowledge. In addition, many studies involve the KOLs as hired consultants to the medical device industry. These individuals are viewed as leaders within their respective clinical community and have likely been involved in publishing clinical studies on techniques, device comparisons, and best practices. A KOL or advanced users seamlessly perform a difficult technique while others, who are less skilled or experienced, have challenges. Thus, a common goal of a CI study is to determine the behaviors and methods that enable KOLs perform with ease. Lastly, should the research team have knowledge of authorship and techniques as published by a study participant, it is a surefire means to gaining a relationship of trust and deepening the conversation.

2.2.2 Conduct Secondary Research

The foundation of advancing clinical practice is through published clinical trials and research. Thoroughly researching the existing landscape by conducting secondary research consists of gathering information collected and synthesized from existing data can add additional insights into user behaviors, opinions, and bias.

This information can be obtained by reviewing journal articles, research papers, other publications, books, state-of-the-art product manuals, procedure protocols, and available demographic and usage data. Conducting a literature review and synthesizing the results into key insights that can be used in the CI study can be time consuming. It requires adequate resources and attention in order to be effective.

Secondary research can be summarized in well-documented reviews with full citations of sources. But it can also be summarized and communicated in shared visual summaries. This prework can be vital as it may provide the framework for the protocol, field guides, and analysis. These documents can serve as a reference throughout the process.

Common suggested topics of secondary research consist of the following:

- Surgical or procedural manuals and clinical reference texts
- Anatomy and physiology
- Competitive product literature
  - Differences of features that create value or hinder usability
- Possible technologies that can be incorporated into a new product

2.3 KICK-OFF MEETING

With any medical device design research program, clear communication between those directly involved in the research and the design is paramount. This frames the research objectives and determines the overall process. Conducting a CI study can be challenging in many regards; however, the planning phase can assist in opening conversations on specific study needs and starting off the research in a positive direction. A kick-off meeting is typically the start of broad communication across the entire development team.

Meeting attendees may include the entire cross-functional development team, the anticipated CI research team, as well as the company leadership. A focus of the meeting is to clarify the goals, objectives, and specific product development targets, if any. For example, in some instances a device company may be looking at what opportunities lay within a market they are not currently involved with and as such are looking for a broad CI study in order to jump-start their process. In this instance, the research team should inquire about what types of devices they would consider developing rather than setting up the study to be exhaustive within one market. In focusing the study further, the results will be more applicable to the target direction of the company. Additional areas of discussion may include existing voice of customer data, preliminary literature reviews along with summaries of any secondary research undertaken or to be undertaken. Lastly, the overall research plan including sample size, strategy of stakeholder involvement, recruiting participants, protocol development, and deliverables should be planned at this meeting.

Specific discussion items:

- The purpose and desired outcome of the research
- Existing research within the organization or in other sources
- The composition of the research team
- The site(s) and contexts
- The study participants: work group or users/consumers to be observed
- The work/process/procedures to be observed

Strength of multidisciplinary teams is the breadth of knowledge and experience each team member has to offer; this meeting enables the entire team to tap into its own resources. Many times the different functional disciplines are unaware of the information that is considered "common knowledge" to the other disciplines. It is imperative that the team members share their knowledge, expectations, and objectives before moving forward. Each functional team member should be asked to define his or her goals and objectives. A well-designed protocol will address the goals and objectives of the greater team not just one discipline or senior management.

2.4 DETERMINE APPROPRIATE TARGET SITES AND PARTICIPANTS (INCLUSION CRITERIA)

Appropriate target sties and participant selection are based on the research objectives balanced with ease of access and the overall timing of the study. There are no set rules for determining sites and participants rather the recognition that a CI study can be a cornerstone of device development and therefore selection should be carefully considered.

2.4.1 Sample Size

Unlike usability studies with published standards and draft guidance from FDA (International Electrotechnical Commission, 2007; US FDA, 2011), there is no magic number for the number of participants to validate a CI study. Typically sample sizes for CI are relatively small by comparison to other types of user-based research. A general rule is 5–10 sites for each separate category of site: home, type of hospital, care center, etc. However, even 2–3 sites per category will typically provide useful data (AAMI, 2014). Ultimately, the number of sites, procedures, and participants in the study are balanced by budget, time, and validity requirements.

Several sampling strategies may be used. One is to attempt to create a representative sample. For example, if 80% of the users are surgical units in large urban hospitals, then 80% of your sites should be surgical units in large urban hospitals. Another approach is to concentrate on "worse case scenarios"—users and sites expected to be the source of the most errors. The logic here is that if you design for the "worse case" it will be safe and usable in all environments. The third strategy is for the sample to include examples of each type of user/site that is fundamentally different from the others, regardless of the relative frequencies. If an objective of the study is to determine the delta between sites or participant types, attention should be placed on having an equal number of each.

Table 2.1 Inclusion and site evaluation criteria

Environment	User	Device or Procedure
Inclusion criteria		
Physical location: • Requires travel • Convenience Type: • Hospital – Academic – Private – Governmental – International • Home • Care center • Specialist office	Demographics Years in practice Type: • Primary device user[a] • Secondary device user[b] Experience in using device Experience in procedure	Target device use Competitive device use Comparative device use
Evaluation criteria		
1.	Overall friendliness and willingness to participate	
2.	Requirements (credentials) to enter and the ability to meet requirements	
3.	Costs associated	
4.	Acceptability to the team	

[a]Primary users are those who directly interact with a device such as a surgeon or a patient using a self-administering drug device.
[b]Secondary users are those who retrieve, prepare, or maintain a device. For example, a circulating nurse or a technician is fundamental for proper device use; however, they are not involved in using it directly.

2.4.2 Clinical Site and Potential Participant Determination Process

Medical device use environments are varied according to the type of device, intended user, and lifestyle. Many devices have multiple locations and types of users and require a broad approach whereas others are highly specialized. A research team can develop a matrix of inclusion criteria as summarized in Table 2.1. This includes details regarding the proposed environment, the device user as well as the device or procedure to be studied. The process of developing a matrix is listed in Table 2.1.

The process for determining ideal locations for CI studies includes the following steps:

1. Create a list of potential use environments

Determine the ideal location/s of device use, for example, hospital or home seeking those places of known connection either by the sponsoring device company, through personal relationships, or previous research.

2. Evaluate use environments

   There are always pros and cons to any environment of use that can assist or prohibit conducting a CI study. Evaluating each site based on frequency of target clinical procedure, anticipated challenges of gaining access, anticipated challenges in the execution of the clinical procedure, and comfort level of the team conducting the research in a specific environment should be taken into consideration. Each environment can be rated for each attribute that may assist in making final selection for study participation.

3. Select environments based on frequency of the target procedure, anticipated challenges at entryand during device use, and relative desire of the research team to visit

   - Frequency of clinical procedure

     Use each of the reimbursement codes, such as DRG (Diagnostic Related Group), CPT (Current Procedure Terminology), and ICD-10 (International Statistical Classification of Diseases and Related Health Problems), for the purposes of billing and providing detailed information. It can be requested that the hospital administrative contact query these codes in order to meet possible inclusion criteria based on the frequency of clinical procedures performed in their hospital or care center. This is not necessarily a definitive approach as frequencies are always difficult to predict, and the clinical calendar is very fluid and subject to rapid change. This information can give the research team a realistic idea of the timing required for a study to be conducted at a particular site. One hardship with this approach is that there may be multiple reimbursement codes for a procedure. For example, the codes for hernia repair include roughly over 50 ICD codes, 40 CPT codes, 10 DRG codes plus modifiers. The research team will need to rely on those professionals who are familiar with the codes and device use. Hospital administration will appreciate narrowing down the codes to a select few rather than running a broad query. The team should expect to receive information that confirms *X* number of *Y*'s was completed during *Z* amount of time. Therefore, adding more accurate predictions of the overall time require fieldwork.

   - Challenges (perceived or real) in gaining access, completing requisite paperwork for entry, and overall friendliness

     For access, many hospitals require training and credentialing by the research team. American Operating Nurse Association (AORN), the Vendor Credentialing Service (VCS), and RepTrax are a few services that can assist in gaining the requirements for being on-site. At minimum, a CI research team should be prepared to provide tuberculosis test results and (for the United States, Health Insurance Portability and Accountability Act) HIPPA training. Seasonally, a flu shot may also be requested; however, this is typically volunteered and not required. The research team should keep their credentials from these organizations on hand during field visits.

There is a unique culture within various hospitals and care centers. The overall friendliness and willingness to participate in the study is very subjective and almost always dependent directly on personal relationships. This is critical to the success of the study, as conducting a CI study requires trust on behalf of the research team with the participants. Sales associates who frequent various locations are typically the best source in providing a thorough understanding of which sites are appropriate and would welcome study participation. They may be able to provide contacts for approvals and scheduling.

The sales team is the frontline on the relationship with users and the device companies. They rely on the relationships as well as the science of the device in order to earn a living. It is possible they are hesitant to enable a CI research team, as they may not have a clear understanding of the intent and purpose of a CI study. Furthermore, they may request attendance during the visit. While introductions are perfectly fine, having a sales representative present during a CI field visit can potentially be a hindrance to free conversation and critique of the device. If the sales representative must attend in order for the visit to happen, they should be encouraged to silently observe the research or be an active participant by assisting with video.

- Challenges in the device use

  No one can predict what will happen in a given clinical procedure. Asking clinical providers their permission to observe and record them while conducting their care requires trust and sensitivity. Many situations can be very grave and serious. A fact of the matter is that the situations we ask to observe may be too circumstantial or hotly debated and will not grant the necessary approvals to be present when a specific high-risk procedure is being conducted. The people involved in delivering care may be just learning how to use a specific deviceand the environment may be too crowded or simply too difficult to access.

In addition, the procedure or device use process itself may be very personal or confidential to the user/patient that observing and recording itself may be deemed inappropriate according to local laws. An example of this is sexual wellness devices.

In any CI study, users will have challenges in device use that a research team hopes to uncover in order to improve device design. Some of these challenges are a result of device design, whereas others are just inherent to the context. Sites can be evaluated on anticipated challenges in device user based on the site and the appropriateness of observing.

- Acceptability to research team

Do you actually want to go to the site? Will the research team feel comfortable observing and carrying out their research responsibilities. Some sites may require significant travel or be situations in which merely observing would make the research team feel too embarrassed, too emotional, or too much left behind on their conscience. In these instances, alternative sites or approaches are required in order to conduct the study.

The process for determining ideal participants for CI studies includes the following steps:

1. Create a list of potential users

Participants should be the people who will actually use or will use the current or future device, system, or procedure. Users may be defined as anyone who touches the device or simply the main user for the delivery of care. There are many different stakeholders in device use and many are involved in the purchase decision. Stakeholders may be defined as users and those individuals who may not be the primary users of a device but assist in the preparation of device use or the patient. An example is the surgeon who is responsible for the case and is responsible for using the device directly on the patient, is the primary user, whereas those who assist in preparing the device or accessing targeted tissue are secondary users. Users may also be those who routinely service medical equipment or those persons who reprocess tools (clean and sterilize).

All users have relative demographics according to the number of years in practice, anthropometry, and openness in opinion. Simply asking experience and measuring body sizes can answer key demographics; however, determining openness may be more of a challenge and may not be able to be answered until the user is interviewed.

Regardless, it is important to list all of the potential users by job description, title, or name and then evaluating potential study participants based on the following aspects: study focus, diversity, and frequency.

2. Select based on the following criteria: match with target user group, diversity, and frequency of conducting target procedure/device use

- Focus

  Depending on the goals of the study, it may be appropriate to focus the users participating to those within a specific market segment, such as novice users or those users who only use a particular device, thereby focusing the study and assuring the device in question will be observed.

- Diversity

  Conversely, different work practices or professional expertise can yield further insights. Having diversity in the study participants enable the research team to compare various user groups and uncover the differences between them. Often advanced users have developed key compensatory behaviors that novices struggle to achieve. Novices who are currently in training, for example, a fellow, with the trainer present can result in the most fertile research-gathering experiences as the trainer often explains all actions as they are undertaken and the trainee often asks questions that the research team has not identified but can yield good insight.

- Frequency

Many users have tried several different devices prior to selecting a preferred device. Likewise, some devices or approaches have never been tried or have been tried, but its been awhile since a user has performed the procedure. It is important for the research team to understand the user preference and experiential bias of a user.

The process for determining ideal medical device uses or procedures for CI studies includes the following steps:

1. Create a list of potential use cases or procedures

In some instances the potential use cases are very straight forward as there is only a handful of patients or approaches that require a particular device. Whereas in others it can be complex, as there are many varieties and nuances of the same condition.

2. Select based on the following criteria:

- Procedure or devices being used
- Frequency—How often is the procedure or work flow performed?
- What is the length of the procedures/work flows?

Selecting and recruiting sites is an ongoing process and can require more time than planned. Once the ideal criteria for participation are agreed upon, the team should begin immediately making contact and getting the process of access started. A fact of the matter is that within the job descriptions of healthcare providers it does not include assisting product development. Explanation of purpose is required even for the friendliest institution or partner.

2.5 INTERNAL REVIEW BOARD REQUIREMENTS FOR CI

An Internal Review Board (IRB) is responsible for the protection of human subjects participating in research; it assures that the rights and welfare of subjects are protected. Their protection is a shared responsibility of the investigators, key research personnel, and the institution wherein the research takes place. There are typically published standard operating procedures that ensure research is conducted ethically and in compliance with federal, state, and local regulations. Any Ethics Review Committees outside the USA (OUS) are equivalent to the IRB.

The determination of whether an IRB approval is required for a CI study is mixed. Some institutions require an IRB approval in order to have any type of study to be conducted within their organization while others merely need to review the study protocol and informed consent forms. Some may argue that any systematic observation of human behavior is a human subject research and as such should undergo review: that perspective is an over generalization but some institutions take that perspective. If there is any question, get the review; it will open doors and make the study more like other studies, which happen in that clinical environment. A reality is that during recruitment, the question of IRB requirements should be asked and then addressed.

Having IRB approval or review for a CI study can only assist in the overall research process and access to clinical sites; however, it requires advanced planning. In addition, there can be fees associated with submission to an IRB and the IRB calendar can meet infrequently potentially causing delay.

2.6 PROTOCOL DEVELOPMENT

A CI study for medical devices has a similar research design as those found in clinical trials and in contextual design principles. CI studies in MDD have structure like a clinical trial, just not nearly as extensive nor as rigorously adhered too. Contextual design is a process that collects data about users in the field, interprets and consolidates the information, and then generates concepts for prototyping (Holtzblatt and Beyer, 2014). By merging the methods, CI studies adapt to the rigor of other clinical practices while providing rich information for design. Adopting contextual design principles into a more formal format enables thorough analysis and ultimately better understanding of the goals and objectives in a CI study by the participants and healthcare administration. For example, many clinical sites are familiar with studies of process improvement. A CI study for the purposes of product design can be viewed as product improvement. While this may not exactly be the intent, it does make the process more understandable by healthcare professionals who have no experience working with product development organizations. Writing a protocol is the first step in bridging the cultures and bringing transparency to the research.

A protocol is a document that describes the study, background and rationale, objectives, methods, and overall organization of the study and is agreed upon by the entire team. In writing a protocol, the team can clarify their thoughts about all aspects of the study. This can help assure the entire team has a shared vision and set course for the study. It is an essential component of an IRB submission and is standard practice in qualitative research.

The biggest difference between a clinical study and a CI study is the potential incorporation of design concepts (low fidelity representations) or conversations on potential uses of technology within the study. Depending on the purpose of the CI study, it may be necessary to follow a specific line of inquiry regarding an idea; these should be incorporated into the protocol to allow broad opinion gathering. The means of concept/idea introduction can be either positively or negatively impacted depending on the presentation means. All concepts or ideas should be presented as neutrally as possible with sufficient explanation to assure communication. For example, if a concept is presented solely in conversation, it is difficult for the study participants to fully understand what the concept really looks like or how it might work. Their vision may be entirely different than what is intended by the research team. Moreover using prototypes to explore an idea or device concept can assure the conversation remain with critique and improvement rather than the generation of new intellectual property. As with intellectual property generation there may need to be additional contracts in place regarding ownership.

Lastly, it is likely that not everyone involved in new product development will participate in the CI study; thereby they will be reliant on the results of the study in order to make design decisions that maximize the overall acceptance and usability of the device. As such, the research team must be diligent with all data.

2.6.1 Contents of a Protocol

Listed below are details regarding each section of the research protocol and templates for ease in study execution. While there is no right or wrong means to organize a protocol, if planning to submit a protocol to a review board, attention should be paid on their specific requirements. Each section is described in detail below.

2.6.2 Background and Overview

This section describes the reasons for conducting the research and includes a brief description of current knowledge. There should be a well-communicated mission that forms the basis of the study. It should also describe the product development goals as a result of the study and any specific details that should have study focus. In some instances a brief description of the methods used and anticipated deliverables may be included.

2.6.3 Objectives

Based on the discussion during the kick-off meeting and secondary research, the team should develop a list of potential objectives. The objectives or goals of the research are broad statements of what is planned. They should be simple and specific with brief explanations.

2.6.4 User Profile or Inclusion/Exclusion Criteria

This section describes target study participant qualification criteria for the purposes of recruitment. It should describe the level of experience based on volume of cases and the acceptance criteria for the diversity of device brand selection. In addition, the target sample size based on project schedule, budget, and availability should be discussed. The attributes of both inclusion and exclusion should be clearly communicated.

2.6.5 Methods for Data Collection

The methodology section is the most important section of the protocol. It should include detailed descriptions of how the CI study will be undertaken. It should include the minimum and maximum number of in-field researchers in order to maintain an uninhibited environment for the study participants. In addition, the number of video cameras, their respective set-up, and the timing of placement should be described. A detailed description of consenting forms, identification of responsibilities, data handling including any post-processing such as data syncing with simultaneous video capture should be provided.

The protocol should have a description of ethical considerations relating to the study. This may include how or from whom ethics approval will be sought, informed consent, and any specific practices undertaken. For example, this might be the avoidance of recording patient-specific information in order to conform to HIPAA requirements. In every study in healthcare, informed consent must be provided. This may be written or provided orally; regardless, consent should be noted in the collected data.

Lastly, a copy of the field guides may be included. Field guides are described in detail below and are intended to be supporting datasheets that assist in collecting specific metrics. These serve as a reference while on-site and as a reminder to the research team that their identity and purpose should be communicated immediately.

2.6.6 Study Materials Required

A list of equipment required should be included in the protocol. Prior to the site visits, the target devices and/or the competitive devices may be requested by the research team in order to explore the use functions of the controls and gain a thorough understanding of the device in question. Almost always the list for field preparations includes the following items: digital still camera, video camera, audio recorder, tripod, field guides or data collection sheets, protocol, and notepads. Other items may include extra batteries or specific tools for data collection.

2.6.7 Data Management Plan

This should discuss the handling of data at the end of a site visit and include details as to where it is storedand the length of time it is stored. Because the CI study involves patient information as well as performance information by clinical providers, data management and security is a priority. Where the data are to be maintained, the process of downloading the data and preparing for the next site visit should be included. Finally, the length of time the information will be kept should be identified. A participating site may request this information prior to approval.

2.6.8 Target Schedule

The target schedule includes all site visits and their locations, and is often updated in real time throughout the study as site visit schedules change rapidly. Overall program management can become challenging if the schedule of site visits is too tight and travel is involved, or in contrast, the clinical procedure schedule is delayed. An accepted mantra in clinical trials is that the perfect way to cure a disease is to study it: the same is true with device studies.

2.6.9 Protocol Template Example

Below are examples of protocol templates and a full protocol prepared for submission for IRB review (see Figure 2.2).

A large extended CI study project plan is listed below. This broad study on interventional neuro-diagnostic angiograms was conducted over the course of two years. The study plan is described in detail below and was used for IRB submission and ultimately approved.

2.6.10 Toward Rational Design of the Human–Device Interface in Catheter-Based Interventions: A CI Study

Unmet Need

Systematic methods for descriptive analysis of physician hand movements during surgical or catheter-based interventions have not been previously reported. Although such data would provide critical guidance in the design of surgical tools, there is currently little to no information about the chronological and spatial integration of operator movements during catheter-based interventions, minimally invasive surgery or open surgery. The hand itself can be conceptualized as a device with more than 20 degrees of freedom. The essential function of this haptic unit is to provide physical coupling between the cognitive process and the environment, translating intention into action. The ideal surgical tool is a contiguous extension of the haptic unit (the hand) that enables an expanded range of effector actions and environmental effects. In reality, however, there is an interface between the hand and the surgical tool and this barrier can

1. Protocol Title
background

2. Study Goals
list study goals

3. Study Materials Needed
may include: current product offerings, prototypes, competitive products, etc.

4. Objectives
1. map the activity
   - *~ define the work groups or activities*
   - *~ define use patterns*
   - *~ define challenges*
   - *~ define design parameters*
2. characterize user profiles and use environment
3. characterize use of device
4. characterize documentation in electronic medical record
5. characterize reimbursement, if any
   - *~hospital (sites) and users*

5. Users & Observation Schedule
list how many observations, where they are to be held, with whom, and observing what types of activities or procedures

observation Schedule- Not specific to the exact date and time but general information on how many observations each week/day. Schedule is dependent on availability of clinical staff and clinical calendar.

6. Methodology
the number of outside observers and/or design team members attending an observation will be no more than 2-3 in order to minimize intrusion to clinical environment. Patient and clinical staff consent will be the responsibility of the design team. Video and photography equipment will be managed by the team to simultaneously capture activity throughout device use.

7. Data Management
7.1 collect data
- ~ collect data using field guide (notebook or app)

7.2 prepare data
- ~ review notes after each observation. clarify data and fill in the gaps that may have been missed due to time limitations or other constraints in the field
- ~ all notes must be transcribed after a field observation into a word document that can be shared with the team. If photography is required to support observations, visual documentation should be included in the document.

7.3 sharing data
- ~ the output of the field guide notebook is a single word document containing notes from a single observation.
  the output of the ipad field notes app is a flattened pdf document containing notes from a single observation
- ~ file naming should follow this convention:
  - YYYY.MM.DD* FACILITY UNIT FIELD_NOTES.INITIALS.doc
  - YYYY.MM.DD* FACILITY UNIT FIELD_NOTES.INITIALS.pdf
  - * this reflect date of observation

7.4 upload data
- ~ upload data to server in appropriate folder
- ~ use file path: \projects\CI STUDY\facility.YYYY.MM.DD

8. Schedule

1-2 weeks	kick-off
	determine appropriate target sites and participants
	write protocol
1 week	prepare for field
4-6 weeks	fieldwork: pilot, site visits
2-4 weeks	data analysis
2 weeks	develop insights
	prepare data visualization/s

Figure 2.2 Protocol template.

introduce variable levels of interference and impedance between the cognitive process and the intended task. The objective of this project is to model surgical effector outputs as a function of operator psychomotor inputs using synchronized multimodal image data recorded during clinical cases of transcatheter cardiac and neurological intervention. The models and multimodal image database generated by this project will be used in the development of new device opportunities as well as interactive educational tools to train physicians to perform these advanced applications.

Project Description

Goal: The goal of this project is to characterize surgical effector outputs as a function of operator psychomotor inputs using synchronized multimodal image data recorded during clinical cases of transcatheter endovascular interventions. The models and video databases will be used to identify new product development avenues and to develop teaching tools for physician trainees.

Research Plan: We plan to use several multimodal video channels to create a detailed record of cardiac and neurological endovascular procedures. This video will synchronously document physician hand movement, surgical effector output as represented by the fluoroscopic image monitor, and spatial orientation feedback represented by rotational movements of the plane of fluoroscopic imaging with respect to patient anatomical axes. Video data will be time stamped to enable analysis of functional coupling and interdependency between channels. Video presentations created from this data will also allow trainees to correlate specific manual operations with specific effects in the surgical environment in the context of the operator's spatial orientation and view (described below). This will allow us to discretely describe each procedure through a detailed quantitative task analysis. After each video data collection session, there will be a post procedural interview with the physician to document the upstream cognitive process that formed the intent underlying each surgical action.

Currently the most effective devices to measure hand movement are electromechanical or magnetic sensing. These work on the hand and measure kinematic parameters at various predetermined locations of the hand and digits. Unfortunately, these methods profoundly restrict natural hand motion. As our data collection cannot disrupt patient care, we plan to use small visibly conspicuous, fiduciary markers positioned on the surgeon's gloved hand. The fiduciary markers will enable us to determine specific joint angles and hand positions in discrete time-stamped video frames. While this is not ideal, it will afford a crude description of joint angles and position for each hand. Once the multichannel video has been created data will be analyzed for the following:

1. End-effector results from specific directional hand movements
   a. One vs. two hands
   b. Identification of factors that affect end-effector performance
   c. Biomechanics of user–tool interactions
2. Task analysis of procedure/s
   a. Efficiency of tasks
   b. Cognitive models of decision-making
   c. Vigilance requirements throughout procedure

Milestones

Our first goal will include the development of a multimodal, multichannel video recording system to be utilized within the catheterization laboratory. Recordings of clinical cases will be used to produce training modules. These modules will comprise three video feeds: operator hand movements, fluoroscopic image monitor display, and imaging plane position in relation to patient anatomical axes (pending new hardware). The display of all the video/animation is synchronized to provide the observer with a clear understanding of the discrete manual operation required to produce a specific effect in the surgical environment (Figure 2.3).

The user interface presents options for basic video operations using Java Foundation Class controls, ActiveX controls, and so on in a browser interface. Options include the ability to launch a desired window to full screen and change video frame rate. This delivery platform provides a high degree of scalability and portability with one central (or mirrored) storage methodology.

Using the system developed, we will expand our research to compare cardiology techniques to neurological techniques. Our goal is to understand the subtle working nuances for each discipline. In addition, we will also compare fellows to attendings (novice–expert) to better understand the learning curve of proficient endovascular therapies.

Resources

Investigators with significant expertise in the human factors of medical devices specifically with regard to hand tool design. Extensive clinical knowledge—Center for Surgical Innovation (UC) and Radiology (UC). Extensive experience in clinical data capture and integration—Center for

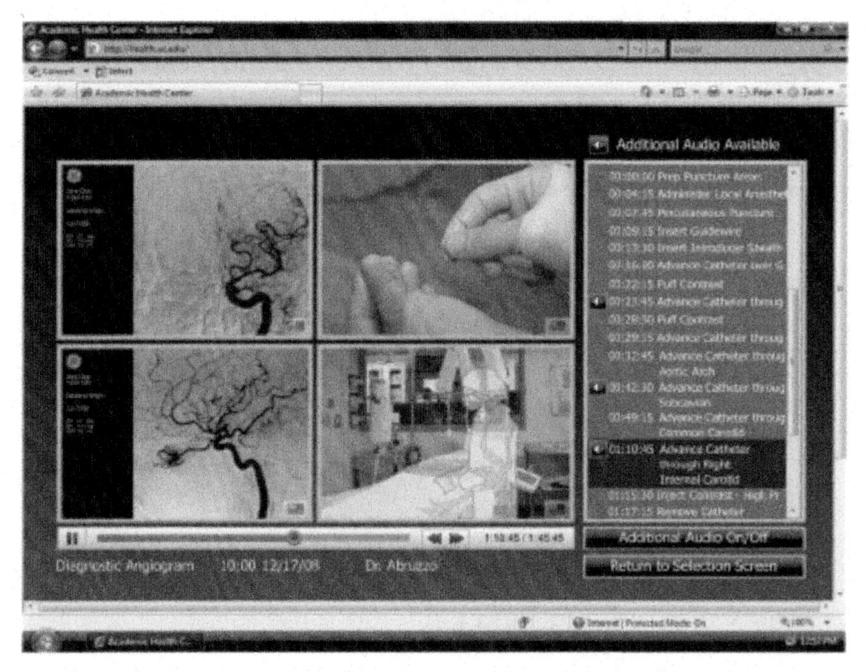

Figure 2.3 Video synchronizing system user interface.

Surgical Innovation (UC). The team includes the following: faculty, graduate students, and undergraduate co-ops in Biomedical Engineering and Industrial Design (Table 2.2).

2.7 CONDUCTING A CI STUDY OUS

The majority of medical devices are designed with the intent to be used globally. As such, the user requirements of those from other cultures should be considered as early in the design process as possible. By conducting an initial CI study in the USA, the target questions and observations can be explored and the protocols tightened prior to taking on a global study. There must be attention placed on fundamental assumptions made by the design team regarding cultural differences. These assumptions should be proven/disproven using evidence gathered during the study.

Studies conducted OUS often rely on a translator, and as a result, additional attention and detail is placed on the research protocol to enable comparison with US data. If budget permits, pilot study OUS in 1–2

Table 2.2 Diagnostic Angiogram Case Study Example of a Study Overview
CI Study Overview: Diagnostic Angiogram

Subjects	*N* = 50 procedures
Inclusion/Exclusion Criteria	Eligibility will be determined by the physician responsible by interview at the time of the patient's procedure. There is no intended gender, age, racial, or ethnic distribution. There are no specific exclusion criteria from this study
Study Design	Nonrandomized study design. The study population will be obtained from current and future patients in radiology, neurosurgery, and cardiology
Study Devices	All tools used internally throughout procedure
Site Visits	*1 Procedure*: Eligibility assessment, informed consent, confidentiality agreement
Visit Summary	Study visit is approximately 30 minutes to 2 hours: – introductions and welcome to clinical staff – complete confidentiality agreement and consent – set up video equipment – record procedure – remove video equipment
Additional Information	This study intends to passively record routine procedures. There are no additional risks involved with this study. All patient data will be de-identified in post procedure processing
Compensation	NA

hospitals and assure translation is correct, assure cultural awareness is maintained and fully appreciated by team.

Specific recommendations:

- Start the study in USA
- Clearly identify protocol, details, and nuances captured within data
- Seek access approval immediately: at the onset of the study
- Large sample sizes are not necessary; however, variety does help provide richer information
- Take a small, efficient, well-trained team
- Explore OUS culture with representatives currently working there (sales, hospital administrations, etc.)

Experience

- Sales team can be a good source for initial contact for access approval
- Approval to use photography is more difficult than simply observing

2.8 BEST PRACTICES

- Create templates
  - Common questions for kick-off meeting
  - Protocol format
  - IRB submission forms
- Keep a list of clinical sites visited and participants
  - Include key contact information
  - Overall experience in having them involved in a study
- Expect uncertainty
  - Confirmations and cancellations may be last minute
- Complete secondary research as thorough as possible: this informs everything from the budget required to complete a study to individual questions
- Get credentialed through RepTrax, VCE, etc.
- Present ideas visually, neutrally, and not in passing conversation
- If IRB is required, seek collaboration with them immediately and ask what is typically required for study approval

REFERENCES

AAMI, 2014. AAMI TIR51:2014 Human factors engineering – Guidance for contextual inquiry.

Beyer, H., Holtzblatt, K., 1998. Contextual Design: Defining Customer-Centered Systems. Morgan Kaufmann, San Diego, CA.

Holtzblatt, K., Beyer, H.R. 2014. Contextual Design. The Encyclopedia of Human-Computer Interaction, 2nd ed.

International Electrotechnical Commission, 2007. IEC 62366:2007 – Medical devices—Application of usability engineering to medical devices.

US FDA. 2011. Guidance Document: Applying Human Factors and Usability Engineering to Optimize Medical Device Design. US FDA, Silver Springs, CO, USA.

CHAPTER 3

Contextual Inquiry Methods

Mary Beth Privitera
University of Cincinnati and Know Why Design, LLC, Cincinnati, Ohio, USA

Contents

3.1 CONTEXTUAL INQUIRY METHODS

3.1.1 Overview

The essence of contextual inquiry (CI) is to include focused observation and possibly having a conversation while a user is performing a task of interest, gathering artifacts, and taking field notes, while conversing with the user in a seemingly informal manner. This requires the research team to be adequately prepared, travel to the workplace, follow the appropriate requirements for access, gain approvals for digital recording, and bring along all the necessary equipment. Having a plan to collect quality data is necessary, especially in medical device development where schedules are quite busy and at times uninterruptible.

M.B. Privitera: Contextual Inquiry for Medical Device Design.
DOI: http://dx.doi.org/10.1016/B978-0-12-801852-1.00003-4

It is a reality that medical devices are used at all hours of the day; an observation may happen in the middle of the night and in a remote location with limited resources. This chapter focuses on presenting various techniques that have been used in the field for discovery of user requirements regarding a medical device. The purpose of the chapter is to present methods in order to demonstrate best practices and the adaptability of a CI study. The quality of data collected is largely dependent on the team in the field. Those in the field are the ones asking the questions and aiming the camera. To maximize their experience, specific methods may be necessary to elicit the true user need and/or the causality of events.

At first glance it may seem relatively simple to just watch people and talk to them and it is. However, doing this for the purposes of data collection requires specific approaches and interview skills in order to capture the most important information in a CI study. This chapter focuses on the following observational approaches: simulated reflection, role-playing, fly on the wall, and think aloud approach. Specific measures in observation are described in detail. Additionally, interview approaches—structured, semi-structured, narrative, and unstructured—are described in detail with the most important interview tip of listening carefully.

3.2 APPROACHES TO OBSERVATION

CI observation is always obtrusive and the research team has an obligation to respect the rights, needs, and desires of study participants. Sensitive information will likely be disclosed and could be of particular concern for everyone involved (Creswell, 2014). Similarly, with any observational research the Hawthorne effect must be taken into consideration. The Hawthorne effect essentially describes an effect that is unavoidable by virtue of the research being conducted (Wickstrom and Bendix, 2000). For example, despite best efforts to remain distant and unobtrusive when observing, the researcher's presence does change the dynamic of the environment especially when people know they are being studied or observed. The term originates from a study of worker productivity in response to lighting conditions. It was only when the study was being conducted that productivity increased; therefore, it was concluded that the research itself was responsible for the change (Diaper, 1990). In medical device development, there may be a tendency for users to follow rules/protocol more closely when being observed. They may even have slow deliberate movements that seem unnatural. This behavior is valuable and can assist in recognizing that the presence of an observer is influencing behavior;

this is especially true if the user completing the task has minimal experience. This can be an advantage when during analysis it is determined that despite best practices, by device design, the device is a challenge to use even under the tightest use protocols.

While in the field and during specific observations, the team should expect surprises and contradictions. Contradictions happen in the form of what participants say is not what they do or in discussion they state a perceived fact then after more conversation they state the opposite. This will be the norm. However, the research process should enable focus, build relationships through non-verbal and verbal communication as well as seek explanation for attributes that are not understood or assumed despite issues encountered. It is the responsibility of the research team to seek true answers, opinions, and clinical implications. The need to observe others and process the information gathered is the cornerstone of a CI study (Holtzblatt and Beyer, 1998). In watching closely, an understanding of the environment of use, social interactions with other stakeholders, devices, tools, and user methods can be analyzed. This means it is important to both watch and listen.

The practice of medicine is highly social and involves many different people with different backgrounds and priorities. The goal is to determine the true value and interactions these users have with devices throughout a completed task. All observations must be conducted at the site of the user in the environment that they typically operate. Selecting an approach to observation is dependent largely on the timing and willingness of participants. In some instances, timing will require an abbreviated study while in others the participants will be highly interactive and enable to design research team to participate using simulated means. For example, most studies are completed within a given time frame in order to maximize impact on the product development projects currently being worked on or being planned. To meet a deadline, the observation may need to include simulated experiences. Admittedly simulation is not reality; however, walking through a simulation with users in the environment is better at gathering information than assuming the product development team has all the use information or guessing. Note, there is no right or wrong choice in selecting an approach to observation, only better or worse.

Below is a list of considerations when selecting an observational approach:

- If observing a procedure, where is the procedure performed most frequently? (Use Environments)

- What devices or procedures are required to be observed in order to meet the research objectives? (Research Competitive Products and Alternative Techniques/Procedures)
- What are the lengths of time for the procedures/work flows? (Schedule)
- Evaluate the ease of conducting CI during the proposed context.
- Is it appropriate to be in the room during the procedure?
- Should the procedure be recorded remotely and then reviewed in context in an Interview Format?
- What is the level of "stress" in the environment when the procedure is performed?
- How large is the Team—external—and how many team members can participate in the CI data collection process?
- How comfortable will the participants or the research team be in context/onsite in use environments? If observing in context is too extreme other CI methods should be considered.

3.2.1 Overt Observation Approach

Overt observation happens when the research team clearly identifies themselves and the purpose of the observation prior to the start (Figure 3.1).

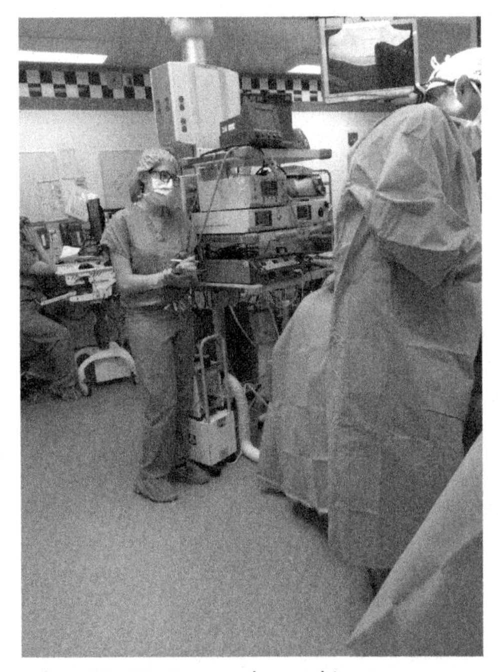

Figure 3.1 CI researcher, Erin Wenig, overtly watching surgery.

The team may be immersed in the clinical environment for entire shifts or partial shifts and will generally have approval from the user and those that they interact with for conducting the research, that is, everyone is aware that the research is being conducted. In this instance, there is an element of observing behavior without participating in it. This approach is useful when the tasks being carried out by the users require concentration or is warranted by the situation. For example, a research team may want to understand the procedures carried out by novices or resident physicians. The novice may request that they not be interrupted during the case or it may be obvious that interrupting is just a bad idea. Another example is that of a surgical procedure that has unanticipated complications. In some instances these are easy to discern: excessive bleeding or direct comments by the surgeon. In others, they may be subtler and it might just be a sensation of stress and quiet that is going on at the time. In this approach, the interactions with the user can be limited. The researcher may only be allowed to interject questions during breaks or pauses in task activity.

For example, at some point in every CI study the research team will need to have an appreciation for being silent. In Figure 3.1, a CI researcher is observing a surgical case and taking notes. In this particular case, the surgeon was very experienced but had not performed the surgery being observed in awhile. In fact, the surgeon was hesitant to allow an observational study given the situation. Approaching the hesitancy of the surgeon with respect and patience along with clear communication of study objectives enabled the research team to gain surgeon confidence and build a relationship throughout the case. Once trust was gained, questions to the surgeon were welcomed.

This approach works well when the following conditions are present:

- When interrupting a clinical case to ask questions would be considered a disruption in care.
- If observing newly trained providers.
- If the situation during a clinical case gets tense. This is the default position of researchers while conducting a CI study.

Overt observation is limited in that while it assists in the identification of specific tasks and the overall timing of the tasks, it does not assist in gaining an understanding of why the tasks were completed in a particular way. It does not necessarily provide causality nor capture opinion. However, nearly all CI studies for medical device development use this approach at some point during field visits.

3.2.2 Think Aloud Approach

The think aloud method consists of asking people to think and talk about their actions while solving a problem or completing a task and then analyzing the resulting the responses (Solomon, 1995; Jaspers et al., 2004). This approach is the most common and preferred approach in conducting a CI study. It is based on a master and apprentice model (Beyer and Holzblatt, 1999); the research team takes on the role of the apprentice and learns from the master. In this approach, users are encouraged to verbalize what they are doing and thinking as they complete a task. This reveals aspects of their opinions and behaviors that would be difficult to discern based on observation alone. This approach relies on a common request, "Can you please describe what you are doing while you are doing it?" The ability to ask questions, as a procedure is ongoing provides further explanation, as often users are not aware of everything they do.

For example, conducting a CI study in the interventional catheter laboratories (Figure 3.2) can pose significant challenges for observation as the majority of device–patient interaction happens in the operatory and the researchers are commonly asked to observe in the non-sterile control room. In these procedures, catheters are introduced (most commonly) in the femoral artery and then directed into the heart or brain guided by the use of fluoroscopy. It is common for CI researchers to be bound to the control room to limit their exposure to X-rays. The control room is a room located just off the operatory and is used for overall procedure recording. As a result of location of the device (inside the patient) and the researcher (insider the control room) a think aloud protocol is virtually the only way to understand what is going on with the patient unless the research team has considerable experience and training observing these procedures. To truly understand without asking, the

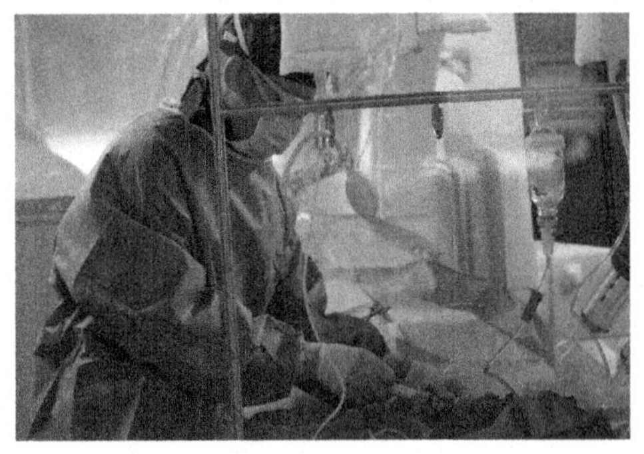

Figure 3.2 Think aloud approach used in the interventional catheter laboratories.

team would need a fundamental understanding of anatomical identification from fluoroscopy images. Having the clinicians verbalize their steps as they went through them enabled a clear understanding of the procedural steps, ability to ask questions concerning clinical decisions, device use selection, and modifications made to the device. In this particular study, the clinicians routinely spend time modifying the catheter tip in order to better navigate the arterial system.

There are two common types of think aloud approaches:

1. Concurrent discussion, wherein the users discuss as they conduct a procedure as a stream of consciousness. The research team should alternate between watching and probing (Beyer and Holzblatt, 1999).
2. Retrospective discussion, wherein users are asked to comment on their processes as they watch a replay of their behaviors or refer back to a specific incident observed (Van den Haak et al., 2003).

The think aloud approach can be used for all CI studies. It enables a direct observation of tasks and explanation simultaneously and uncovers the opinions and challenges as verbalized by the user. Ideally, this approach is completed in real time; if unable to accomplish this, it is best to target specific events within an observation and then conduct a retrospective discussion. In this approach, researchers may need to remind the participant repeatedly to articulate their feelings and opinions as they may get focused on the complexity of the clinical task. For example, if a cardiac surgeon, who is performing cardiac bypass surgery, where typically the saphenous vein is removed and utilized as the bypass graft, has gone along in the beginning of the procedure explaining every step and now pauses to examine the harvested vein prior to implant, they may need a gentle reminder to start explaining again. Finally, the use of this approach relies on the personality of the participant. Participants can be shy or have a particular affinity for a particular part of the procedure that overrides the discussion. As such, gentle probing may be necessary to redirect or restart the conversation.

This approach works particularly well and very naturally when observing training sessions. Here, the trainer will go over the underlying theory and importance of tasks while the trainee may ask questions that may be novel to the research team thus providing rich data collection. By the same token, the research team should always expect the experience level of the user to correlate with their explanations; the more experienced, the deeper the explanation.

Lastly, in procedures where the patient is awake, providers may be reluctant to provide concurrent think aloud discussion as they may want the patient to be distracted from what is going on with the procedure and this would draw attention to it. Before the procedure begins, the research team should ask permission to ask questions and have the procedure explained to avoid any uncomfortable issues.

3.2.3 Use Simulation with Reflection

This approach involves the user demonstrating their processes, techniques, and tool uses within the environment of use while the researcher asks questions and seeks explanation. In this approach users perform a procedure or part of a procedure using anatomical models and the actual tools. This approach is best completed with adequate time allotted to enable further explanation and discussion. It immerses the research team in the procedure and questions can be asked in detail as time pressures as well as critical life measures are not existent. During the simulation the clinical provider should follow every step in the procedure including donning personal protective wear (Figure 3.3). This is especially important if a goal of the study is to explore ergonomics and the physical interaction of the hands and tool as gloved hands, dry or wet, provide different tactile feedback that may be important to capture.

This approach is best at the start of a CI study. It allows the research team to become familiar with the clinical procedure, the tools, and the methods currently being used. Similarly, it works when the deadline is tight and there is not enough target procedures or study enrollment potential for a given clinical condition. For example, in the words of Dr. Art Pancioli, Chief of Emergency Medicine at the University of Cincinnati, "The best cure for disease is to study it!" In clinical practice if it is nearly impossible to schedule the procedure or catch it as it occurs,

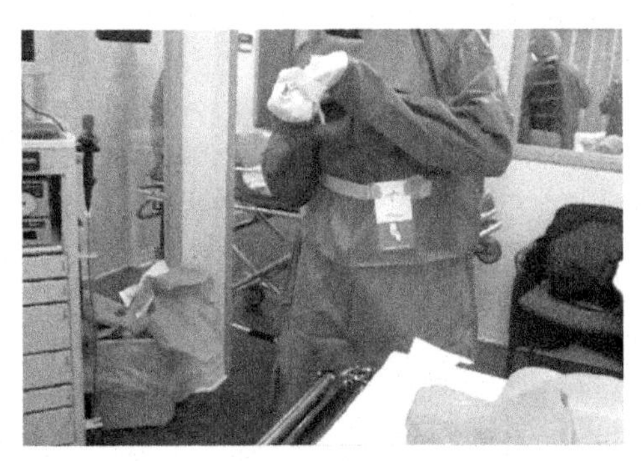

Figure 3.3 Provider simulating placement of central lines using torso mannequin.

simulation with reflection is very appropriate. For example, it would be very difficult to study the diagnosis and treatment of ischemic stroke (blood clot in the brain) unless the research team was integrated within the clinical department and likely partnering with their clinical research team. In this instance, having an actor or simulation mannequin stand in for patient presentation with a full team of clinical providers can yield insight into the realities of their care.

There are a few considerations to be made aware of in this observational approach. These can include the lack of variability in patient presentation and time assessments will be inaccurate. Because this approach requires the use of mannequins or anatomical models, the nuances of patient presentation are not necessarily accurate. Providers must always deal with multiple variables within their care, as no two patients are exactly alike. Using mannequins, all participants have the exact same patient unless pre-programmed to mix conditions. This can be a benefit if comparing tools but a deficit if the goal is broad understanding of patient anatomy and physiology. The final limitation is using the approach to assess time. Due to the lack of a real patient in this approach, there is no real sense of urgency. While certainly the providers can act as they would, there are typically a few small details missing that can add to the overall time spent on a procedure. For example, many simulations inject surrogate fluid as injections: the clinical team often overlooks the time it takes to order a specific medication, retrieve, prepare, and inject. Another example is the need to go find equipment. In simulations, everyone plans and will make sure all equipment necessary is available, whereas in practice there may be makeshift assemblies used to accommodate missing or broken items.

To inform the research team at the onset of a central line study, a simulated reflection approach was undertaken (see Figure 3.4). The goal of this interactive session was to walk slowly through the procedure and enable the provider adequate time to explain the processes and techniques taken in the placement of this invasive line. For this particular case, an experienced practitioner who routinely trains residents was recruited. This enabled the research team to take on the role of apprentice and experience the procedure in detail. While in reality, to place a central line an experienced practitioner can complete this within 5–10 minutes once all supplies are pulled and personal protection is donned: this session lasted an hour. The timing is directly tied to the amount of questions generated by the research protocol, new questions, or clarity questions and the willingness of the provider to assist in the research team's learning.

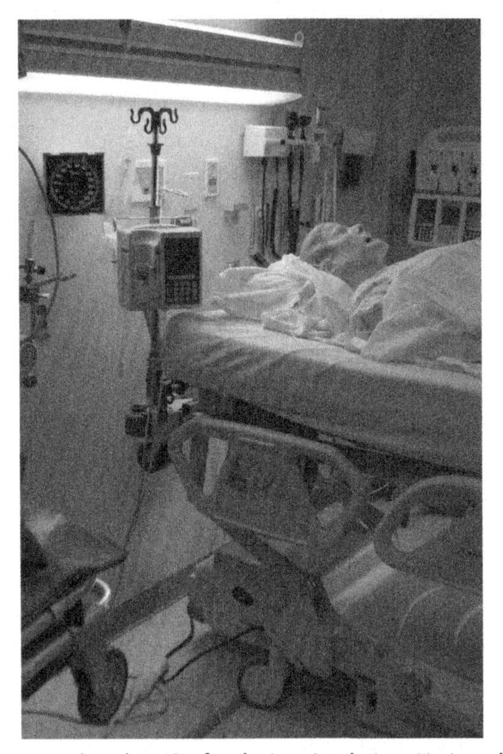

Figure 3.4 Mannequin placed in ICU for device simulation. He is a sleepy patient.

3.2.4 Role-Playing Approach

Role playing is a technique commonly used by researchers studying interpersonal behavior in which researchers assign research participants to particular roles and instruct those participants to act as if a set of conditions were true (Lewis-Beck et al., 2004). In medical device development, this requires very open-minded personnel. In this approach the research team interacts directly with the user and the device being studied (Figure 3.5).

The role of the researcher can be from the perspective of the patient or the provider. In many instances role-playing is conducted during animal laboratories where exploration of the design problem and the mitigating factors is the focus of the exploration. In this case, the physician can demonstrate a particular skill and then a member of the research team can attempt the same skill. This requires an anatomical model—mannequin, animal, or cadaveric. Having the research team perform the roles while specific conditions and circumstances are being explored

Figure 3.5 Interactive role-playing laboratory wherein research and development team from Bard Medical interact with the device and animal model under the guidance of an expert surgeon, Dr. Krish Gaitonde, in order to explore endourology tools.

Figure 3.6 Dr. Gaitonde illustrates anatomical structures while explaining device use.

enables the CI study to explore potential areas that may or may not have been identified at the onset of the study. In addition, it directly builds empathy for the user, their challenges, and can bring out specific usability requirements that can inform credible solutions.

As mentioned earlier, this approach requires a willing provider to enable the environment, the tools, and the time to be available. If using a laboratory, this can take the better part of a day as lecture and illustrations by the surgeon expert is assured (Figure 3.6).

It is best completed anytime in the development cycle; however, for CI specifically, it helps at the onset or middle of a study. This approach can further inform interview and observational protocols as well as insights. Finally, it is the perfect tool to use if there is a specific idea or innovation that needs to be explored in context for increased understanding.

This approach is conducted more readily by those who teach residents or who have considerable experience explaining their methods in practicing medicine. Admittedly, the approach is a completely manufactured

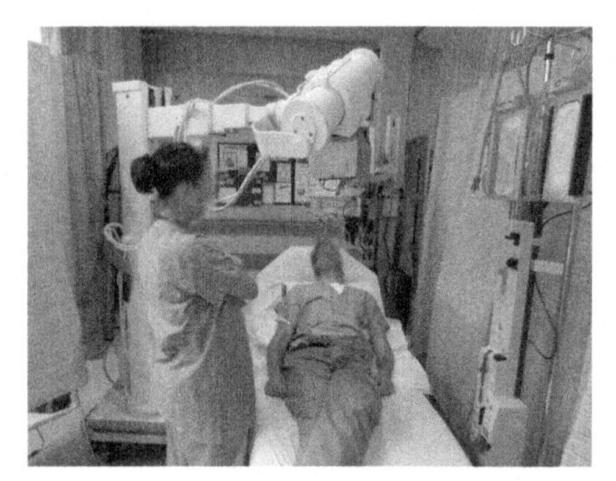

Figure 3.7 Researcher role-playing a patient in the UCMC Center for Emergency Care.

situation and cannot create the physical adaptations, professional skills, and emotional states that are experienced in reality by practicing providers. However, the team will appreciate the challenges and difficulties faced by the providers more readily first-hand.

Another example of role play in a simplified manner is in Figure 3.7, wherein a research team member acts like a patient in order to understand the challenges of angulation relative to patient positioning when taking X-rays in the emergency department. The provider explains the typical positioning of the bed and X-ray plate along with the importance of understanding the angle at which the X-ray was taken clinically. Through this role-playing simulation, the research team was able to generate empathy for the challenges consistent with X-ray angles and their relationship to the provider's ability to decipher and diagnose from the image.

3.2.5 Note Taking: Specific CI Observational Measures to Consider

Often a CI research team only gets a limited opportunity to record actual measurements while in the field. Certainly there is a wealth of information collected on the video, however dependent on the detail necessary in the study; accurate note taking may be necessary to fully make sense of all the activities within a clinical procedure. In other words, there are elements that a research team may consider actually listing or measuring. These include the following:

Environment—Describe the lighting conditions, the temperature of the room/s, the overall layout, a description of any shared or individual spaces, and the overall impression of the arena of device use.

Task—Describe the discreet steps a user took from a particular onset of the procedure through to a specific completion, for example, from the time the patient enters the operating room until they leave. This includes specific actions and processes undertaken by the users or team of providers.

Interactions of people and devices—Describe the nature of routine and/or special circumstances that require different interactions between people and the devices they use.

Devices—Describe in detail the specific devices used in a procedure, how they were used, for what intended purposes, and the overall impression of device use. For example, describe if the device was used as intended based on corporate knowledge, training materials, and previous VOC data. Focus on any compensatory behaviors or work-around/s that users have developed in order to complete their procedures.

Artifacts or documents—Artifacts are physical items that are found to assist users in completing a task. For example, it is important to gather or record items that are found that may remind users of a particular condition, situation, or record their procedure.

Users—Describe their attitude for the day and how this changes/does not change as the observation progresses; describe their role with the larger organization, their values, and biases.

Patient—presentation, anatomy, and response to clinical procedure.

Taking notes during the observation can contribute to data analysis and serve as prompts during debrief sessions.

3.3 INTERVIEWS

For the purposes of a CI study, an interview is a semi-structured conversation. In a conversation each party has the freedom to explore new topics and dive a little deeper into one area based on choice. For a CI study to enable comparison across several participants or observations, the interviews are semi-structured. The researcher conducting the interview does not rigidly follow an interview outline rather uses a guide to assure all questions are asked. Interviews are conducted throughout the study and can be both informal as the team prepares for the data collection and formal in structure during data collection in the field. The formality in interviews during data collection is only in the use of interview field guides.

Interviews, at the beginning of a CI study when the research team is trying to initially learn about the user and the use context, can be very informative and can assist in setting up a final protocol. Likewise, interviews conducted after some data analysis is completed can assure accuracy in translation and interpretation. For example, in learning about a complex procedure such as mitral valve replacement. Upon completion of the secondary research the team may want to conduct a preliminary interview to get more familiar with the anatomy and process. In addition, this is the perfect opportunity to practice speaking the language of medicine and will increase comfort, ultimately the ease at which subsequent interviews occur.

It is important to note the CI interview differs from formal structured interview techniques moving from one question to the next without variance. In CI, the interview guide enables the research team gain specific knowledge while jumping around freely much like a conversation with a colleague. By conducting them in a free but rigorous manner, the research team can ensure reliability and validity of the data collected. This ensures that the findings confidently reflect what the research set out to answer rather than any personal bias that may exist on the research or product development team involved in the CI study.

Interview protocols or guides should fully anticipate the conversation and language used to focus the inquiry on specific goals of the research. In medical device development, as mentioned, a research team must learn medical terminology and get comfortable referring back to it. For example, anatomically the left atrial appendage (LAA) is a flap of heart tissue that bends back on the left atrium (one of the fours chambers of the heart); this piece of anatomy should be referred to as the LAA and not "the bendy thing that folds back on to the top of the heart." By using the correct medical terminology, participants in the CI study will have respect, immediately recognize the seriousness in preparation for clinical understanding, and thus have a deeper, possibly more clinical and scientific discussion.

In any CI study, there are always instances where the day is just not a good day for the participant and the answers provided are truncated due to circumstances beyond the control of the researcher. For example, a trauma surgeon who has been up all night and has had clinical challenges during a CI study observation may not be in the best of moods at the end of their case. It may be better to schedule an interview on another day. A technique to determine participant mood, if not already apparent, is to simply ask or use a tool (Figure 3.8) to discover the participants' overall frame of mind.

The mood of the interviewee participating in the study directly affects where and when the interview takes place. Discretion is with the researcher to know when to request an interview and when to walk away. Figure 3.9 highlights the most common moods of clinical personnel working in the interventional neuro-radiology suite during a study.

Figure 3.8 Circle one: a tool to inquire about participant mood.

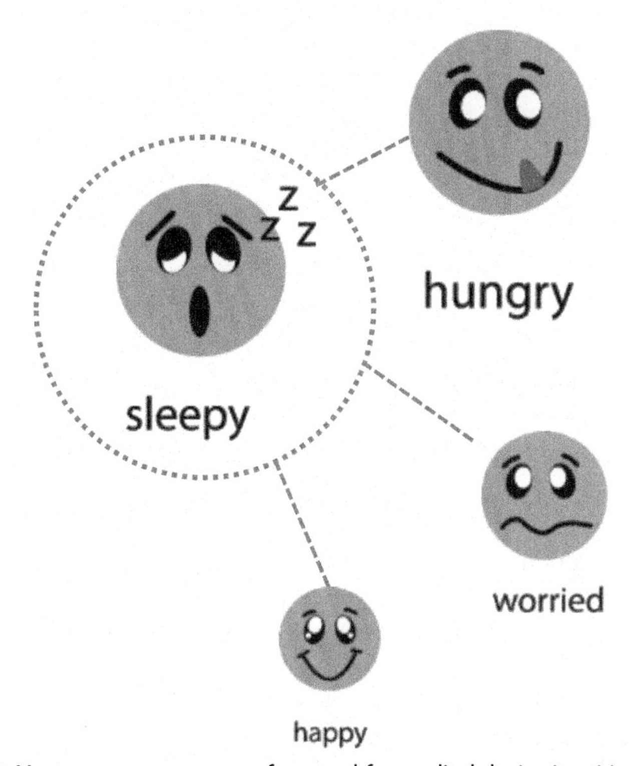

Figure 3.9 Most common responses for mood for medical device inquiries.

This methodology will result in furthering relationships with study participants. The sustained contact between interviewer and interviewee over a period of time extends the possibilities for interactive production of knowledge. The focus is always on the participant's job and the clinical science they practice; thus, CI is never a sales call, it is an information exchange and learning experience. Additionally, conducting a CI study provides medical device manufacturers an additional means of interacting with their customers and as such can lead to long-lasting collaborations.

3.3.1 Interviewing Approaches

There are three primary approaches of interviews in CI studies: semi-structured, unstructured, and narrative. It does not use a structured approach, wherein the interview is designed to learn specific information in predetermined focus areas and asked in a rigid predetermined schedule. Rather the semi-structured, unstructured, and narrative interview techniques promote natural conversation, which is imperative in a CI study. All interview approaches are useful at different times in the process. For example, the unstructured interview is useful at the beginning of the study when the team must quickly learn the anatomy, the devices, and the current procedures. Semi-structured and narrative approaches are valid approaches at any time in the study. Each can yield new discovery and prompt design information.

3.3.2 Semi-Structured Approach

A semi-structured approach promotes the development of data through interaction. This approach is defined by a flexible and fluid structure in which the interviewee has the opportunity to jump from topic to topic with ease (Bechhofer and Paterson, 2012; Lewis-Beck et al., 2004). It is most useful when the research team knows something about the topic being discussed (Wood, 1997). This type of interview is completed with a diligent approach and ensures that all topics are discussed. The structure of the interview is usually organized around an interview aide or interview guide (Creswell, 2014). This guide contains the topics, themes, or areas to be covered during the course of the interview, rather than a sequenced script of standardized questions. The aim is usually to ensure flexibility in how and in what sequence questions are asked, and in how particular areas might be followed up and developed with different questions.

It is based on the generation of knowledge through interactions of discussion wherein the interviewer and the interviewee have an active, reflexive, and constitutive role in the process of developing knowledge (Creswell, 2014; Lewis-Beck et al., 2004). Moreover, it enables the use of probes and tools in order for the interviewer to further explore the discussion topics (Figure 3.10). Through interactive discussion, the causality

Figure 3.10 Clinical providers using an opinion probe wherein different colors of post-it notes represent varying perceptions.

in opinion can be expressed. That is the "why do you think (that) is so?" while indicating a particular part of the device/procedure within a natural flow of conversation. Thus, the data are derived from the interaction rather than simply the answers provided by the interviewee.

This open, flexible approach is intended to prompt the interviewees to share their own accounts, perspectives, perceptions, experiences, understandings, interpretations, and interactions as they view them. Figure 3.10 represents users placing color-coded post-it notes based on their perceptions and opinion. This simple tool was used to gather view-points on aesthetic perceptions of design concept visualization.

3.3.3 Unstructured Approach

The unstructured interview allows interviewees as much latitude as possible in answering open-ended questions and going off in directions of their own (Fontana and Frey, 2005). Like the semi-structured interview, the unstructured interview is also seen as an interaction accomplished between the interviewer and the interviewee. The unstructured interview does not rely on closed-ended or structured questions. Rather, the interviewer pur-sues information about a given topic by asking open-ended questions or merely prompting the interviewee. It is more exploratory in nature. The interview may be taped or the interviewer may take notes during the inter-view or, more rarely, shortly after the conclusion of the interview.

An unstructured interview allows flexibility in the design of the interview and the questions that are asked. It is typically used at the onset of a study in order for the research team to explore questions that can be used in a full study. It is rarely used in a full study, as it would garner in consistency in data collection thus making data analysis somewhat problematic.

3.3.4 Narrative Interview

This technique involves using detailed "stories" of experience, not generalized descriptions to elicit a response from the participant. Narratives come in many forms, ranging from descriptive ones that recount specific past events (with clear beginnings, middles, and ends) to narratives that traverse temporal and geographical space—biographical accounts that refer to entire lives or careers (Lyle Duque, 2009). Participants engage in an evolving conversation: narrator and listener/questioner, collaboratively, produce and make meaning of events and experiences that the narrator discusses.

When the interview is viewed as a conversation, rules of everyday conversation apply: turn taking; relevancy; and entrance and exit talk to transition into, and return from, a story world (Lyle Duque, 2009). One story can lead to another: as narrator and questioner/listener negotiate spaces for these extended turns, it helps to explore associations and meanings that might connect several stories. Narrative accounts require longer turns at talk than are typical in "natural" conversation, certainly in mainstream research practice. Therefore, in conducting CI for medical device development, adequate time for extended narration is a challenge.

Narrative interviews can be useful if there is an ill-defined design goal and the team is seeking baseline information. Opening up the research interview to story telling places control of the conversation in that of the study participant. Sometimes, it is next to impossible for a participant to narrate experience without context or tools. These assist in generating the details that may be insignificant to the study participant but important for the design team to realize. For example, there are fundamental tenets, procedures, and processes that have historical legacy that become difficult opinions to break. An example of this is the laryngoscope used daily by emergency medicine physicians and anesthesiologists for tracheal intubations (Figure 3.11). This tool assists in gaining an airway in patients who are not breathing or about to undergo surgery. The function of the device is to enable visualization of the vocal chords and the glottis. The user holds the device in their non-dominant

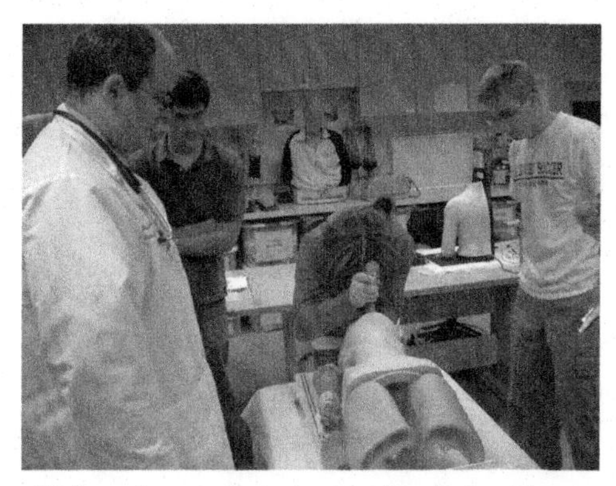

Figure 3.11 Biomedical engineering students role playing the process of intubation using a laryngoscope.

hand pulling up and away from their body while the device is inserted into the mouth of a patient. The user must bend forward in order to gain an adequate view of the vocal chords while they insert an endotracheal tube.

Basic ergonomic analysis of the situation will yield that the design of the laryngoscope is poor for the following reasons: excessive shoulder flexion (upwards) under heavy loads by non-dominant arm, inadequate gripping surface (hand can easily slip off, there are no form features to prevent it), and the requirement to hold a static bent position while performing the task. In narrative interviews with attending physicians, it was discovered that in order for some physicians who have poor shoulder strength, they were encouraged to hold a gallon of milk out in front of them for periods of time daily in order to gain the strength necessary to perform this life-saving procedure. Although there are specific research objectives, such as learning about a particular device, narrative interviewing means following participants down their trails. As a result, genuine discoveries may be uncovered.

Certain kinds of open-ended questions are more likely than others to provide narrative opportunities. For example, asking what happened in the past can yield insight into their level of understanding of their technological and social aspects of their workplace. A question may be "Can you tell me a situation in which you have done X with Y? and how it turned out." This prompts the participant to recall the details, turning points, and other shifts in cognition, emotion, and action. Another prompt can be, "when you are training a novice, can you please share with me any tips or tricks you have found useful? Why or why not?" Additionally, asking for negative situations can also yield interesting insights. In this instance, the question "Can you share an experience by yourself or a colleague where a procedure did not go as planned ... and why?" While ideally there is first-hand story telling, enabling the participant to speak as if they are speaking regarding someone else and it is not their behavior or situation they are referring to. In doing this, anecdotal situations, experiences, and opinions can be further explored. In almost all CI studies in medical device development, a form of narrative questioning will be appropriate. A reality is that in conversation, all people love a good story.

3.3.5 Good Interview Techniques

Being a good interviewer in conducting a CI study means that you are really a good conversationalist.

A good conversationalist asks follow-up questions, particularly if they did not get it quite right the first time. They rarely need to ask mulitple follow up questions as there is decorum in conversation that we move on. Additionally, a conversationalist in CI must have passion for the questions they are asking. Passion itself can be contagious and motivational. By demonstrating passion toward the daily activities of others, more information, practices, and disclosures will occur naturally. This requires asking questions with clarity, sincerity, and enthusiasm.

Likewise, the interviewer should keep the conversation about specific events and experiences with requested examples. They should be prepared to jump around the order of the questions if the participant gets on a roll. This is true even when the topic of the roll is not necessarily the target conversation. In doing this, lasting relationships are built. Of course, the relationship with study participants can be temporary in nature; however, the relationship built through one study can lead to participation on another. Also, the research team may have question that needs to be clarified after data analysis has been conducted. The most important skill of an interviewer or conversationalist is to listen carefully and respond accordingly.

The following techniques describe what to do and what to avoid in conducting interviews as a conversationalist for a CI study. While no study is perfect in every way, interviews are critical in a CI study and assist in determining user opinions.

Establish a Rapport with the Study Participant

Make sure the participant consents and understands the confidentiality of the interview. Participants are allowed to speak off the record in a confidential manner and be provided a full description of what confidentiality means for the study. This aids in developing trust and is particularly important as observing clinical care involves disclosures which can be very personal in nature.

Know what Questions to Ask

Asking questions is relatively straightforward and easy for most researchers; however, knowing what to ask that determines value as communicated by study participants is an art. Interviews should be targeted around several focal questions designed to cover the main aspects of the research question.

Avoid Asking Leading Questions

Leading questions are those questions that subtly prompt the user to answer in a particular way. They can result in false or slanted information and are typically yes/no type questions. The discussion should focus on the how? And why?

Only ask one question at a time and then wait patiently

Asking too many questions or having multiple questions within one statement can be confusing. Keep the interview simple, ask one question then allow the participant time to process the question and answer at their discretion. Many newly trained researchers fail to recognize when a participant does not understand a question and when they are preparing a thoughtful response.

Avoid Questions that are too Open

Questions that are broad in nature may be used to get the conversation going; however, in some instances they are so broad that the conversation can take on expansive possibilities in their answers. This may lead the interview in a completely wrong direct and waste valuable time.

Include Questions of HOW, WHY, WHEN, FOR WHAT PURPOSE

These are the most important type of questions for CI studies. The purpose of the interview is to seek causality of behaviors that describe the context, the process, the social interactions, and the device interactions in the delivery of healthcare. Yes/no type questions should be limited to confirmation of understanding rather than information seeking.

Use Prior Knowledge in the Questions

As previously mentioned, the more the researcher understands about the clinical problem, the anatomy and the current methods employed by clinical practitioners, the more tailored questions can become during an interview. Prior to conducting a CI study researchers should prepare by reading available medical literature, marketing literature, and use manuals provided by device manufacturers. Often the corporate sponsors have considerable reference material regarding the procedures being studied.

Consider Using Tools to Assist the Interview

A tool to uncover the emotions a study participant may have is to give them a bank of images. The images should be vary in nature and as an example, include images of people walking across a tight rope, a fast car, anatomy. The interviewer asks the participants to match the image that matches their feelings with the steps in the procedure (Figure 3.12).

This process requires the participant to think in metaphor. It is helpful and readily accepted by those less trained in medical device use such as techs or some nurses. However, its been a painful experience with some surgeons who are more literal thinkers. For example, thinking about complex medical procedures in metaphor using this technique can be foreign to a "what you see is what you get" physician and they have been somewhat problematic in garnering true answers. In these instances, the surgeon becomes impatient and intolerant of the tool. Using the tool

Figure 3.12 Study participant matching images with procedural steps.

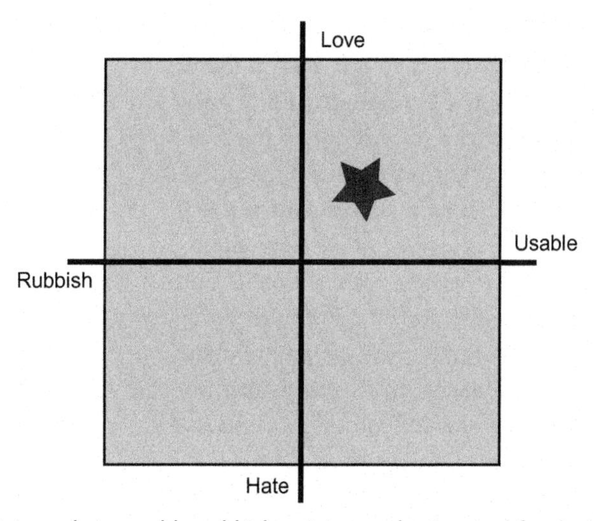

Figure 3.13 Love, hate, usable, rubbish opinion evaluation tool for devices.

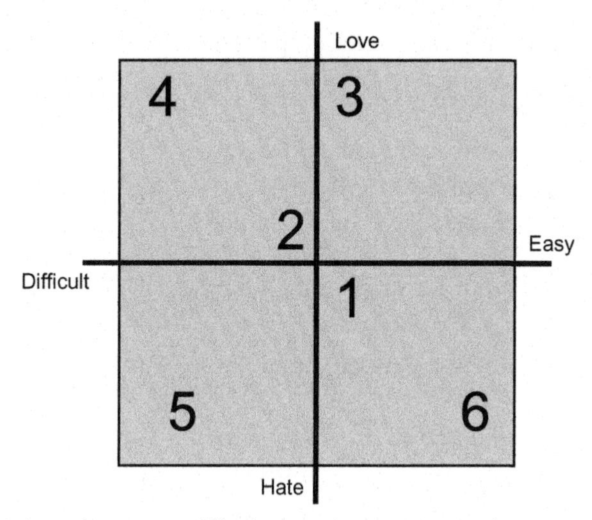

Figure 3.14 Love, hate, easy, difficult procedure opinion tool.

shown in Figure 3.13 of just asking their opinion may be a better option. In Figure 3.13, the horizontal axis plots the perceived usability of a device and the vertical axis plots the like/dislike opinion. In this tool, the term love/hate is used purposefully as users do have strong opinions regarding the tools they use (Hyman and Privitera, 2005).

Similar to the device evaluation, the same tool can be adopted for opinion evaluation of the procedure steps overall (Figure 3.14). In this instance the study participants are asked to plot each step in its corresponding location according to ease of use and overall like/dislike.

In Figure 3.14, step 1 was relatively easy and slightly disliked, whereas step 5 was both difficult and disliked. In contrast, step 4 was both difficult and loved. This response is common in complex procedures. For example, in an interventional neuro-radiology study the step of deploying various treatment devices to the interior carotid arteries (interior of the brain) through a catheter stemming from the femoral artery (upper leg) and maneuvering it in order to complete therapies on intracranial aneurysms (ballooning of an artery in the brain) was the most difficult and challenging part of the procedure. This challenge and the ability to complete it successfully and efficiently was precisely the love element in the procedure. Therefore, making drastic improvements in this step may not be of value to the current customers. A tool like this plots two opinion attributes and recognizes that opinion is not one-dimensional.

3.4 BEST PRACTICES

- Plan an approach for observations; know whether it is acceptable to ask questions or if it is required to remain silent.
- Use illustrations in notes.
- Ask the provider to think aloud as they perform the task. If forgotten or the participant is shy, provide a gentle reminder.
- Practice the medical terminology and then practice it again. It needs to flow easily in conversation.
- Simulation is a good technique to learn both by doing (role playing approach) and as a data collection tool.
- Take great notes: it is difficult to measure an operating room from video.
- Do not be afraid to ask for anything deemed "trash" by study participants. Often they will allow you to take package inserts or instructions for use that would otherwise be discarded while in an observation.
- Listen carefully during interviews and observations.
- Listen carefully.
- Carefully listen.
- After listening, form confirming questions.
- Encourage story telling
- Have a directed and focused conversation with eye contact.
- Given the opportunity to play doctor or nurse while guided by an expert, take it.
- Use tools and probes to elicit deeper responses.

REFERENCES

Bechhofer, F., Paterson, L., 2012. Principles of Research Design in the Social Sciences. Taylor and Francis, Hoboken.

Beyer, H., Holzblatt, K., 1999. Contextual Design: defining custoner-centered systems. Morgan Kaufmann, CA, USA.

Creswell, J.W., 2014. Research Design: Qualitative, Quantitative, and Mixed Methods Approaches. Sage Publications, USA.

Diaper, G., 1990. The Hawthorne effect: a fresh examination. Educ. Stud. 16, 261–267.

Fontana, A., Frey, J.H., 2005. The interview. In: Denzin, N.K., Lincoln, Y.S. (Eds), The SAGE Handbook of Qualitative Research. pp. 695–727.

Holtzblatt, K., Beyer, H.R., 1998. Contextual design. In: Soegaard, M., Dam, R.F. (Eds), The Encyclopedia of Human-Computer Interaction, 2nd Ed. Aarhus: The Interaction Design Foundation.

Hyman, W.A., Privitera, M.B., 2005. Looking good matters in medical device design. Med. Device Diagn. Ind. Mag. 27, 54–63.

Jaspers, M.W.M., Steen, T., van den Bos, C., Geenen, M., et al., 2004. The think aloud method: a guide to user interface design. Int. J. Med. Inform. 73 (11), 781–795.

Lewis-Beck, M.S., Bryman, A., Liao, T.F., 2004. The SAGE encyclopedia of social science research methods.

Lyle Duque, R., 2009. Review: Catherine Kohler Riessman (2008). Narrative Methods for the Human Sciences. Forum: Qualitative Social Research, 11 (1), Art. 19.

Solomon, P., 1995. The think aloud method: a practical guide to modelling cognitive processes. Inf. Process. Manage. 31, 906–907.

Van den Haak, M., De Jong, M., Jan Schellens, P., 2003. Retrospective vs. concurrent think-aloud protocols: testing the usability of an online library catalogue. Behav. Inf. Technol. 22, 339–351.

Wickstrom, G., Bendix, T., 2000. The "Hawthorne effect" – What did the original Hawthorne studies actually show? Scand. J. Work, Environ. Health 26, 363–367.

Wood, L.E., 1997. Semi-structured interviewing for user-centered design. Interactions 4, 48–61.

CHAPTER 4

Executing and Documenting a CI Study

Mary Beth Privitera
University of Cincinnati and Know Why Design, LLC, Cincinnati, Ohio, USA

Contents

M.B. Privitera: Contextual Inquiry for Medical Device Design.
DOI: http://dx.doi.org/10.1016/B978-0-12-801852-1.00004-6

4.1 OVERVIEW

Fieldwork is time-consuming and expensive, developing a systematic plan for fieldwork, which assures the objectives are met as smoothly as possible (Figure 4.1). While visiting users of medical devices there are expectations. These expectations cannot be overlooked, as they may be the fundamental requirements for admission. These may include providing a proof of credentialing, signing in at a specific location within a hospital, appropriate dress, and the recognition that breaks may be delayed or nonexistent.

There are four major key components of this phase: (i) contacting study participants, (ii) developing the field guide/s, (iii) building an observational tool kit, and (iv) developing a data management plan. In addition, there are many small details to be considered, which are discussed in detail below. This step typically takes the longest amount of time due to scheduling constraints.

This chapter provides an overview of what to expect, while conducting a CI study in the healthcare environment. This includes expectations a research team may have as well as those that study participants expect are discussed. In addition, overall program management, key considerations prior to field visits, assigning research team roles, and data collection and management tools are discussed in detail. Specific field etiquette for the operating room, general hospital/intensive care units, emergency care centers, and international sites are described.

4.2 EXPECTATIONS IN THE FIELD

Expectations come from all sides involved in a CI study, including the study sponsor, the research team, the clinical participant, and the hospital or institution in which the study will be held. A sponsor of a study expects to be updated often, to be included in decisions and that the overall study will be executing as efficiently and professionally as possible. The CI research team expects through their preparation and recruitment that the site visits run smoothly and rich data will be collected. In addition, study participants demand professionalism and respect for their work place. Each of these expectations is likely not surprising; however, they should not be taken for granted. Regardless of experience or preparation, there will likely be unforeseen challenges in the field. Some are easily resolved, while others may cause the visit be rescheduled or long wait times by the research team. *Patience and a positive attitude are required.*

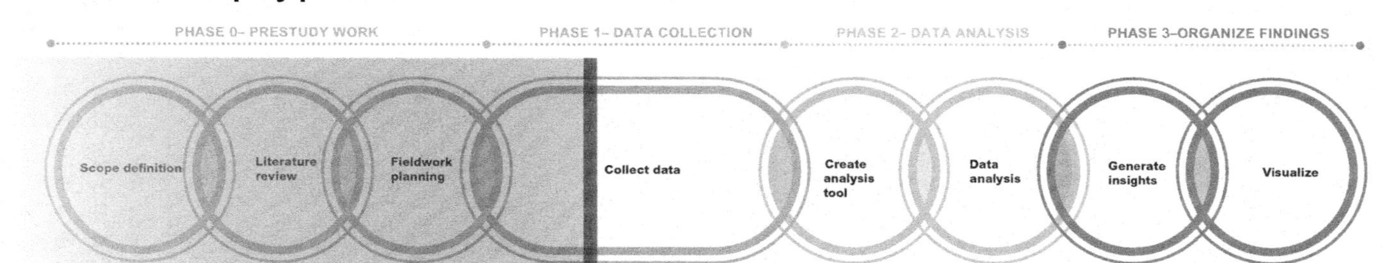

Figure 4.1 Preparation for the field requires making contact with medical professionals, meeting their requirements for admission, developing field guides, and a data management plan.

4.3 ANTICIPATED CHALLENGES AND POTENTIAL SOLUTIONS

While in the field, overcoming the challenges of purpose—*"why are you here?"*—can be readily resolved with a solid, brief explanation of the study goals along with the name of the contact person who has assisted in facilitating the site visit. The research team should be prepared to explain themselves several times throughout the day as shift changes or care sites change and new personnel enter during an observation. Those who are in the hospital leaderships, such as charge nurses, may need reassurance that privacies are being protected and that the research has been approved through appropriate vetting persons/agencies for their institution.

Data collection will include video of observations, recorded interviews, collected artifacts or documents, and personal memos. Gathering these will require transporting equipments, such as cameras and tripods, which will need to be carried, stored, and tended to. Bringing these equipment into care facilities can be a red flag for multiple parties to stop the research team and inquire as to the purposes of the equipment—"*Why do you have that camera?*" While perhaps seemingly unfriendly, in doing so, the care providers are truly protecting the privacy of their patients. With every medical care institution, there are gatekeepers; these are individuals at the site who provide access or permit the research to be done (Creswell, 2014). A brief proposal may be necessary, which explains the purpose of the research, what can be expected on a site visit, and includes information on how the data are managed. Additionally, many of these individuals may not have prior knowledge of a "contextual inquiry" type study and may need further explanation of it. One approach is to compare a CI study with a process improvement study. These are more familiar and common to hospital administration and they often utilize observational methods.

In addition, challenges in the field, such as timing, gaining access to a procedure of interest, and permission/s to document using recording devices, happen often. Many hospitals require visitors who are there for the purpose of conducting research or sales, to sign in at a specific locations, such as the security desk or main desk for the unit: these are not always easy to find (Figure 4.2). This takes additional time that sometimes cannot be avoided. For example, once on site, there may be challenges in getting to the right spot at the right time for an observation. Imagine you are scheduled to observe and record a surgical case that starts

Figure 4.2 Scrub dispensary for vendors in the sub-basement of a hospital.

Figure 4.3 Researchers, like vendors, may have to retrieve scrubs from an alternative location. Many hospitals supply scrubs for their staff directly off the changing facilities and may not allow for external personnel access.

at 7:20 AM and are not permitted to check into the main OR desk until 7:00 AM. You can expect that as you are often required to check in as a vendor, you are required to then find the scrub dispensary (Figure 4.3) and locker rooms. You should expect to change into scrubs that they provide and find your way to both attain the scrubs, the changing facility for non-personnel, and then finally, someone to escort you into the specific OR for your case. In this case, scheduling a visit over two days can benefit the team. If possible, the first day schedule can be dedicated to learning the ropes of the organization so that the majority of the second day can be dedicated to gathering as much data as possible. It is the standard practice for facilities to provide scrubs, especially in going into the operating room. This is to minimize outside elements from getting into these areas.

Another challenge may be excessive waiting as it is often that surgeries run late, as do patients. This is the "hurry up and wait syndrome," wherein the research team changes their overall schedule in order to catch a specific case, and key personnel have communicated that the timing of the procedure will happen immediately. However, upon arrival, the team discovers a delay due to a myriad of possible reasons: surgeon had an emergency, the anesthesia team (secondary team to that being studied) has not arrived, patient is non-cooperative for the case or does not provide consent, and so on. In these instances, the research team should adopt a reaction of patience and understanding. Obvious signs of frustration or negativity will only damage the relationship and possibly jeopardize data collection. As a rule, the relationship with the study participants should always be preserved.

4.4 PROGRAM MANAGEMENT

The overall study schedule will depend on contacting the right users, their timing in conducting the right procedures, and their willingness to participate. The process can never be started soon enough and ample time should be given for uncovering specific requirements clinical sites may have for access. Once the identities of the ideal participants are determined, immediately start checking their availability and scheduling. Many times their schedules will dictate the timing of the observations and interview, as well as possibly the method of your interview. In some instances, it may even require an adjustment in the protocol. To some degree in every CI study, flexibility will be required especially with regard to the schedule.

Scheduling site visits can be a challenge depending on the procedure being observed, if they happen infrequently. The easiest site visits to schedule are those focused on procedures in the operating room. While these clinical calendars are fluid, they are at least scheduled in advance with the exception being trauma. Regardless, the schedules always vary and are dependent on the length and the type of procedure or device use being observed, as well as the content in the study protocol. The team needs to be realistic and allow time before and after each observation as some field visits will run long and there may be travel time between them. In addition, scheduling an observation is easier in some areas of care versus others. For example, scheduling an observation of a trauma or cardiac arrest is impossible. It is cost prohibitive and may be an annoyance to have a research team to camp out in an emergency department, therefore, other means of observational data collection may be warranted (see Case Study in Chapter 13 for a case study focused on using simulation).

It is recommended that interviews be scheduled directly after an observation has been completed, therefore, making them much easier to accommodate. This enables the research team the opportunity to seek clarity on any questions that were not asked in real-time during the observation. As sometimes interrupting or even making a noise is unacceptable. Other reasons to postpone an interview may be that the provider may have other patient responsibilities or the provider may simply refuse, be in an ill mood, and/or request a reschedule. In these instances, the research team should be flexible and accommodating. If there are outstanding observational questions, these can be marked in the memos of the team during a debrief session. While not ideal, but certainly helpful, interviews that are held without an observation should (at minimum) have the tools commonly used in the procedure available for the study participant to use during the discussion. By providing the visual cues of the tools, the study participants are more likely to pick the tools up and demonstrate a particular means of accomplishing a task, as it is difficult to remember the detailed technique without having props. In doing this, the team can still use photography and capture some of the device–user interactions.

4.5 CONSIDERATIONS BEFORE ENTERING THE FIELD

Sometimes, despite best efforts to remember everything necessary to conduct a study, something may get inadvertently forgotten. Gathering a CI toolbox for each study can ease the preparations for field visits. The nature of the study will determine what tools should be included. Listed below are the toolkit contents for consideration.

4.5.1 Study Toolkit Contents

- Digital camera (number required?)
- SD memory cards (with large memory capacity that are empty)
- Extra batteries
- Patient Consent forms
- Credentials and approvals for research team members
- Business cards
- Field protocol

- Click* pens or iPad
- Notebook for personal memos
- Honorarium (if included in study protocol)
- Participant log

4.5.2 Patient Consent Forms and Requirements

Consent for photography is always required for any CI study, especially in medical device development. It must be obtained prior to conducting observations and a reality is that consenting may not be in the researcher's control. Sometimes the site of clinical care or study participant will consent patients or patient families prior to a procedural observation. It may be a requirement of the institution that one of their credentialed employees are the only personnel permitted to gain consent. They may or may not volunteer a copy of the consent to the research team. In this case, it is recommended that a copy of the consent is requested and maintained by the team. The consent form is really the researcher's ticket to photograph.

If a researcher is to personally consent the patient, here is what to be expected: initially the researcher should make introductions, briefly explain the goal of the study, and the treatment of data collected. Of importance, the patient must sign an approved form (Figure 4.4). This form may be required at the time of recruiting sites to assure the care facility or study participants that the best research practices are undertaken for the study. It also reassures participants that the team takes the protection of their personal data seriously. If the patient is unconscious, the researcher may explain the terms of conducting research to a family member and receive consent from them. However, if consent is not granted, do not use any recording devices. Reasons for consent denial may be communicated to the team or they may not. Common reasons for refusal include simply not wanting any images or video taken of them, lack of trust between the team and the participant, and finally the (real or perceived) risk that the images may be used out of context of the study for the purposes of malpractice.

Consenting everyone involved is a priority. Every participant and/or patient has the right to refuse photography or recording of any kind and

*Click pens are recommended as pens with caps increase the likelihood that the cap will fall to the floor and require a researcher to pick it up. While in itself is not all that bad, you may not desire to pick up anything from the floor of a clinical site and carry it with you.

Consent and Release for Photography
Study Title

Date

I,________________________________, grant **COMPANY X (X)**, their subsidiaries, successors, assigns, employees, and agents (**X**) the right to video record my procedure of ____________(date)_____________, (year), and use my photographic likeness as it appears on the video tape. I give **X** permission to reproduce, distribute, publish, exhibit, use and/or transmit the same, or any portion of the same by any means, including the internet or electronic media, for any purpose.

As it is common for an observer to photograph or videotape the use of a medical device, any such activity will be conducted in such a way to avoid the disclosure of my face or any uniquely identifying characteristic.

I understand that I may request cessation of recording or filming at any time and at my direction, the observer(s) will leave my presence and, if directed by me, destroy any record of the observation.

I release **X** from any and all liability to me, which may arise from the exercise of rights grated in this Consent and Release Agreement. This Authorization will not expire on any specific date, unless indicated below:

____

Printed name of person granting release ________________________________

Signature of person granting release ________________________________
(Parent or legal guardian if person granting is a minor)

Address: ________________________

Telephone number: ________________________

Date: ________________________

Figure 4.4 Example patient consent form.

while disappointing, it is the law. If permitted, the study participant may still enable the team to observe the procedure just not to record it. In these instances, diligent note taking and onsite sketching are recommended. Thus, having a researcher with good sketching skills can provide a great deal of information. Sketching the overall room and environment layout and user flow helps to record an undocumented case and can tie notes to video when recording is permitted.

4.5.3 Best Camera and Video Practices

Where a researcher aims a camera and the angle at which the data captured is an initial form of data analysis. This filters out less important views and the researcher is deciding what to capture, what not to capture. As an alternative, it may helpful to have multiple video cameras set up to capture as much information as possible. In this instance, a typical camera set up includes one camera to take an overview of the room and another (or multiples) set on the device use or overall procedure. The overview camera can capture the social interactions and device preparations, while the other camera can focus on specific interactions of the study participant and on the primary user. Multiple cameras may require special permission to set up in advance prior to the arrival of the patient and/or additional research staff in the room. This needs to be clarified thoroughly with the participating site. Surprises are generally unwelcome, especially in the operating room.

Specific tips for the camera include:

- Ensure that the video camera is set to a quality and data ratio that will work for your team and subsequent data analysis.
- For privacy, unless indicated in study protocol, refrain from filming provider or patient faces.
- Use a tripod: for the overview angles and when interviewing. This provides a steady camera and note taking during the procedure.

4.5.4 Honorariums for Participation

Most organizations or participants will desire honorariums for participation in a CI study as their time is valuable and they are opening up their workplace for a somewhat intrusive visit. It is up to the organization, that is, planning the CI study to decide whether compensation is appropriate. Compensation is often a direct payment to the study participant or in a donation to a specific foundation. Honorariums affect the entire study budget and can get costly. In addition, in the United States, the Physician Payments Sunshine Act requires manufacturers of drugs, medical devices, and biologics that participate in US federal health care programs to report certain payments and items of the value given to physicians and hospitals. All manufacturers must submit a report of payments to the Centers for Medicare and Medicaid Services (CMS) on an annual basis. They are required to categorize how the payment was received and provide a reason. As such, the CI research team should keep diligent records on any exchange of funding. CMS audits for fraudulent practices. In participating in a CI study, the researcher is requesting the study participant to provide instruction during device use, often termed preceptorship on payment records.

Honorariums range in cost and can be between $150 and $500 for a surgeon for each case observed and site fees can range $500–$10,000. All fees will need to be negotiated directly with the study participants and include the means of dispensing as well. For example, a site may want all fees to go into an education fund as a donation rather than individual payment to personnel.

4.5.5 Patient and Provider Confidentiality

The research team should assume that the study participant is doing the best they can with the tools they have been provided. The job of a CI research team is to take data collected from the field and analyze it looking for various conditions and situations including inadequacies or deficiencies. This can mean that the study participants may, at times, seem as if they are incompetent or perform badly. Taken out of context, the research itself can be problematic for a study participant. It can inadvertently and negatively affect participant's role within an organization. To protect study participants, the *practices from human subject testing should be adopted*. These include:

- Watch closely in order to maintain the security of the information gathered.
- All participants should be given a unique identifier.
- All identifying elements, faces or marks, such as tattoos, should be hidden.

Ethics for all participants are expected and when in doubt refer back to the institutional review board requirements and/or corporate expectations.

4.5.6 Assigning Team Roles for Conducting Study

Conducting a CI study is a team effort and having assigned roles and responsibilities can make the process run smoother. Listed below are examples:

- *Leader*: responsible for developing rapport with the study participant, following the study protocol, including leading the interview and observation. This person is responsible for the majority of the interactions with the study participants while on site. They may also be responsible for gaining consent.
- *Wingman*: assists with the interview and asks follow-up questions, takes detailed notes, and is responsible for the majority of photography.
- *Observer*: responsible for assisting in photography and taking detailed notes. Typically a silent observer dedicated to capturing data and study oversight.

A minimum of two (the leader and the wingman) should be present for site visits. Independent sits visits become difficult to both interview and capture the data as indicated in most study protocols. It is possible for a study team to rotate rolls within the same study or interchange team members. This interchange is most often the observer role. In having assigned responsibilities, nothing is left to chance and everyone knows exactly what is expected.

4.5.7 Developing a Field Guide and Pilot Study

Protocols often have more information than is absolutely necessary in the field. To strategically take notes and assure completion of the protocol, a field guide can be very helpful. A field guide can be as simple as a list of questions and specific observations that are required or they can be programmed apps with tailored design to the study. An alternative is small bound notebooks that enable sketching and have free pages for individual notes. Field guides serve as reminders to the research team and can assist in focusing the lens of the observations (Figure 4.5). The field guides incorporate collated information and quick reference on anatomy (if needed), the procedure, the discussion questions, and the observational metrics. It is an abbreviated version of the protocol and can be used for all memos.

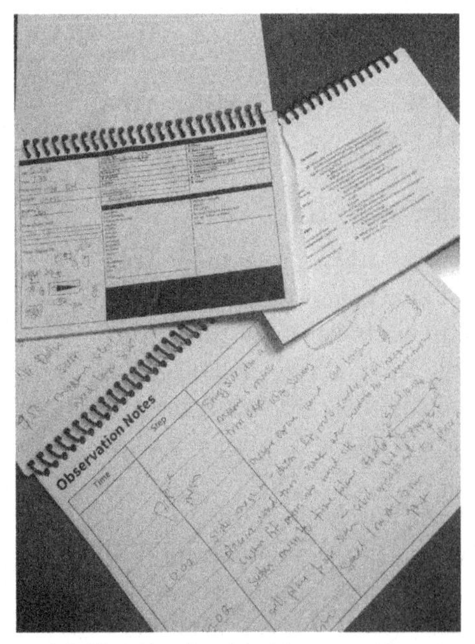

Figure 4.5 Spiral bound field guides for site visit note taking.

Before locking down the field guide/s, an initial run through the entire visit should be completed. This is a pilot site visit and enables the team to assess the field guides and overall process. After completing a pilot visit, the research team should seek any modifications to the field guides before broadly producing and using a final version. Changes and modifications on the fly, while in the field, will cause problems in data analysis ultimately gaining rich information but possibly information that is difficult to make conclusions.

In every CI study, design insights and ideas will be stimulated by events throughout the study. While sharing ideas with participants provides immediate feedback by the user, this should be avoided if not part of the planned protocol. Ideas with merely a verbal description are very difficult to interpret. The vision in the researchers mind may not match that in the participants, thereby perhaps providing false feedback. It is highly likely that the verbally described idea was misinterpreted. For accurate idea evaluation, use concept sketches or another tool to cohesively describe the idea to all of the participants. This can mean modifications, which are necessary to the protocol/field guide. These modifications may include a low fidelity prototype or concept illustration. If communicated without these tools, the loose communication can become a can of worms in data analysis and can provide false evaluation of an idea rather than collecting information that describes what exactly is going on. The importance of integrating this into the protocol or field guides is to prevent inconsistency across site visits and participants. Data analysis and insight generation become problematic and the entire study validity can seem inaccurate.

Listed below are some considerations before developing your field guide. These include:

- Look at the context of the interview. Determine when it is appropriate to hold the interview, before the observation, during the observation, or after the observation. Consider if space is required to conduct the interview. Ideally, if possible, have a pre-interview observation and post-interview debrief to discuss events that you feel were relevant.
- Determine how many people have to be there to observe and record the data efficiently and effectively.
- Determine the tolerance of ideation with the study participants.

4.5.8 Interview or Discussion Guide

Interviews should be conducted as a conversation rather than a line-by-line session of inquiry. It is the responsibility of the interview leader to manage the interview effectively. This means at the onset, remind the study participant that you are seeking their opinions and if they are uncomfortable in answering a question they can decline. In addition, ask if there are any time constraints that must be respected. This can help expedite the line of questioning and the approach. Ideally, the lead researcher should take their time in asking questions. Take care to only ask one question at a time and be patient for the response.

Examples of what to include in a Discussion Guide:

1. The Introduction
   - Thank you for helping us learn more about—Procedure, Product, etc.
   - I am working for X and our goal today is to learn about "Y" and how
   - We can improve the design of "Y" in order to make it easier for you to do your job. You are certainly the expert and we are here to learn from your experience and opinion.
   - During the observation, we will be taking photography and need your consent. Can we take a moment to read over the consent form? *Walk through consent.*
   - Do you have any questions or concerns before we begin?
   - Thanks!
2. Collect Demographic and Background Information

   It is important to collect this information, but this is also your chance to develop rapport with the user. Try not to go through this like a laundry list and try asking few factual questions, and then interject a more personal one.

   Sample questions:
   - How long have you worked at current employer?
   - In current job?
   - How long have you been in the field?
   - Before this, what fields did you work in?
   - What is your proudest moment on the professionally?
   - Out of all of your jobs in the field, what tasks have you enjoyed the most?
   - What is your greatest achievement professionally—awards, hugs, letters from family, raise, promotion, etc.
3. Explore the Primary Tasks/Work
   - Can you please describe your key work responsibilities?
   - Can you take a moment and walk me through the procedure step by step?

4. Procedure/s or Activity Knowledge
   - Can you please walk me through the entire procedure?
   - What are the key decision points in the procedure?
   - What is the patient condition during the various stages/steps. Decision points of the procedure?
   - What level of skill does it take for each stage/step of the procedure? Or alternatively asked, which part of the procedure is the most difficult and why?
   - Do you use special tools or artifacts during the procedure?
   - What other people interact with you during the procedure?
   - Level of enjoyment at each stage—Love/Hate opinion tool, metaphor tool or simple Likert scale.
   - Is special training need to perform the various stages/steps of the procedure? If so, please describe.
   - Can you describe any areas of challenge?—probe for patient position, patient mobility, use of restraints, patient comfort, ergonomics procedure, patient interaction, etc.
   - Please describe areas of enjoyment.
   - Are there any areas (of opportunity) that you feel can be improved? If so, what and how?
   - Do you have any ease of use concerns or challenges? If so, what?
   - What are your other work responsibilities?
   - How do you perform them?
   - How do you know when a task needs to be performed?
   - How do you know when you have done a great job?
   - What other people do you interact with during your work?
5. Product Knowledge
   - What equipment/artifacts do you use (in your job)?
   - How long have you used each?
   - Are there other products you prefer using or have used in the past? If so, what and why?
   - Where did you use them?
   - What is your opinion regarding "Y"?
6. Wrap up

The interview is coming to an end and you want to use this time to make sure you understood the content of the interview, make sure the user does not have anything else to add, and thank the user for their participation and provide closure on what this means. Many times users want to know what are you doing with this information, will you get back to me, etc. The answers to these questions should be discussed with the team and the answers should be consistent between interviewers and users.

- We are about to wrap up is there anything about your work, procedure, situation you want to talk about?
- Do you have any overall comments or suggestions?
- *The leader may engage the wingman at any point, but especially at the end of the interview to ensure all questions were explored.*
- Thank you.

Respond appropriately if compensation questions arise and be sure to clean up any materials as a result of the study. Finally, request the ability to follow-up if you have further questions.

4.5.9 Observation Guide

Listed below are suggested details that can be added to an observation guide. These include:

- Identify tasks in context with the rest of the process
- A task fits a larger story
- Understand how informal collaboration with others fits inside of the formal process
- Look for the larger process, for example, what is happening around the procedure
- Focus on the user's job roles, not their title
- Identify work groups
- Work groups are the core to any kind of work
- No one works alone, what is the social atmosphere like?
- Communities are just big, loose work groups
- Ask your user from whom they are getting task and information
- Remember that work groups are informal and formal

What to capture:

- The roles that the user is playing in the organization
- The user's responsibilities within their roles
- The types of communication that the user engages in
- Evidence of culture: language, tools, music, conventions, and beliefs
- How the user organizes their physical space
- Any artifact that the user uses or refers to
- The user's key tasks, work strategies, and intent

- Breakdowns in the user's work
- What works and does not work in the tools that the user uses

4.5.10 Data Collection Strategies in the Field

The majority of data collected in the field is through the use of video. Knowing what to capture and what not to capture on video, while in the presence of healthcare delivery requires a strategy. Where the camera/s are aimed (Figure 4.6), the amount of information gathered within the frame and the ability to have audio or not have audio provides the research team with great control over the data. Carefully recording procedures can ease data analysis, whereas haphazard videos or videos with missing information can be extremely frustrating in analysis. Aiming the camera should be based on the protocol and specifically focused on those elements of device interaction that can inform design, such as ergonomics.

As mentioned previously, multiple cameras are useful at collecting data capturing both in detail views and broad overviews of a clinical procedure. A tripod (Figure 4.7) or monopod, that is monitored by a research team member is almost always the best means of capturing the overview of the room.

Unless assured the tripod is in a safe location without the possibility of inadvertent falling, it should never be left unattended. There is an alternative to the tripod, that is, to mount the camera on a piece of existing equipment prior to the start of the procedure and use a remote to control the camera (Figure 4.8). For this recording method, the research team needs to have access to the operatory prior to the start of the case, while the rest of the clinical team is also prepping the room. Figure 4.8 shows a

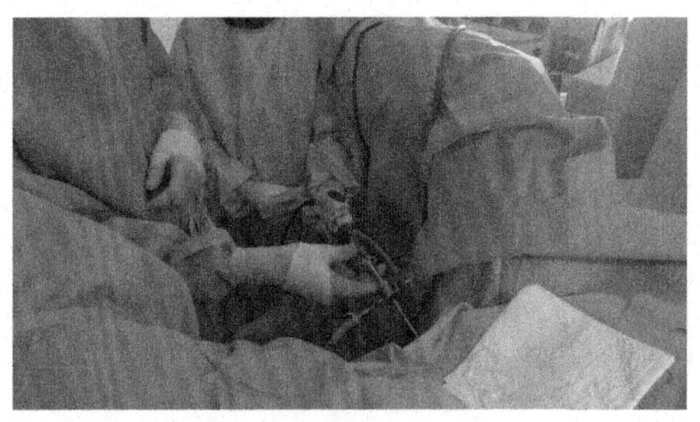

Figure 4.6 Aiming the camera is the first part of data analysis.

Figure 4.7 Overview tripod placed in corner against a wall to protect from inadvertent bumping.

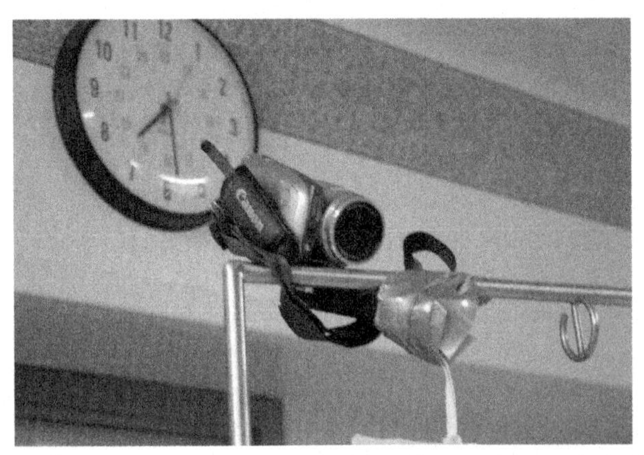

Figure 4.8 Video camera attached to IV support bar across patient table in interventional radiology suite.

video camera that was mounted prior to an endovascular intracranial aneurysm clipping procedure.

Again, multiple cameras and good note-taking can assure that adequate data are captured. One important note is that inevitably once a view is found through a "window" between clinical practitioners that aims the video right on the target data point, it can be expected that a study participant unknowingly blocks the camera view, and the researcher is required to adjust (Figure 4.9). Once adjusted, the cycle repeats itself. In essence, while recording, expect to move from one side to the next

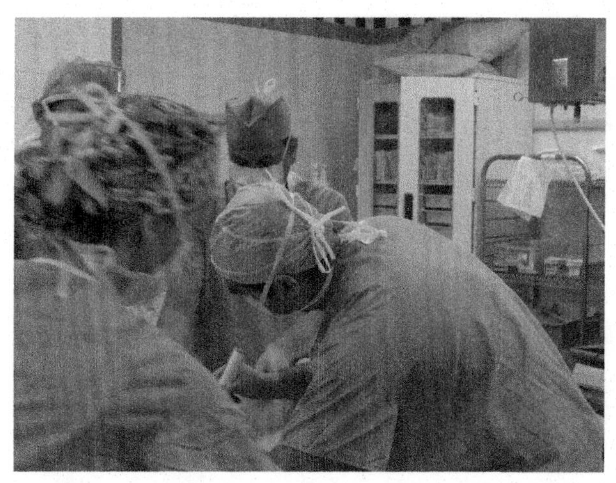

Figure 4.9 Sometimes it is difficult to get an adequate view of what is really going on in an observation. The researcher should rely on overhead cameras or move in this instance.

adjusting the flow of video to maximize the data collected and account for movements of the clinical team.

Overhead cameras can assist in these instances; however, they too suffer the same issue and will happily record the back of a surgeon's head if not adjusted properly. Vigilance is necessary for good photography.

In some cases, it may be required to record both what is happening internal to the patient and what is going on externally with the devices in some minimally invasive procedures. In Figure 4.10, there are two views of patient anatomy that the surgical team used to guide the procedure. On the left is an image generated using fluoroscopy techniques and on the right is that of a scope. From these two images, the surgeon is able to locate where the target lesion is and plan access.

While the majority of data collection will be completed using video, using a still camera has its advantages, such as increased ability to zoom, specifically aim at a targeted event, and acquire higher resolution images (Figure 4.11). The combination of still images and video footage can make a robust data set that can be analyzed using different means.

4.5.11 Silent Observation

In all procedures, there may be times of silent observation. These are times when a provider may be unable to discuss a situation due to privacy or patient condition. Silent observation may be required where the

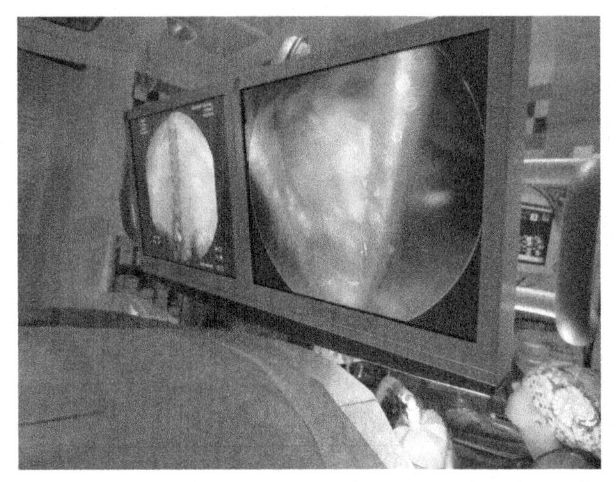

Figure 4.10 In many instances, data have to be captured both inside and outside the patient's body during minimally invasive procedures. This will require multiple cameras and possibly syncing of the data for analysis.

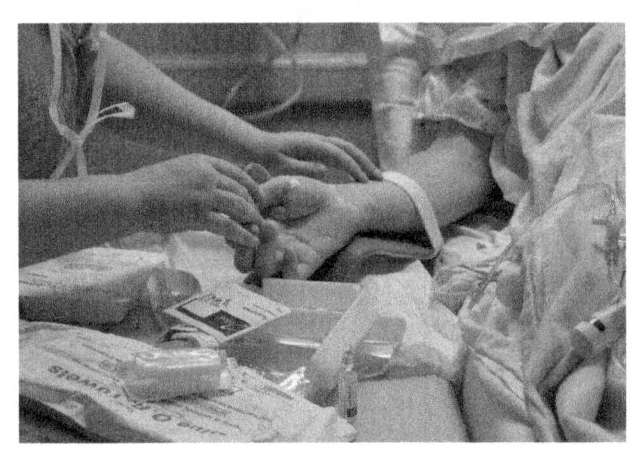

Figure 4.11 Still cameras can capture great detail.

patient is awake and aware. Other times it may happen if a case runs long or significant problems arise. In these instances, it is important to keep the cameras recording. Often these moments can be inquired upon during the post-procedure interview.

4.5.12 Interview

The study participant will almost always determine the location of the interview. This may be a room off to the side of a busy clinical space, it may be in a break room, or it could be in the hall. The research leader should request a private quiet location. The length of time required is directly related to the number of questions and the attitude of the participant. If the participant has just completed a very busy day after night on call it may be best to reschedule the interview to the following day. The discussion should last as long as necessary for the research team to confirm their understanding.

4.5.13 Debrief Plan

A site visit debrief is a casual meeting of the research team at the end of a visit. It is an opportunity for the research team to gather after the observation and further discuss, catch up on notes regarding the visit. It is essentially hitting the pause button on the research and provides time for the team to have dedicated and immediate reflection. In addition, debriefing is used to enhance the accuracy of the account (Creswell, 2014). It enables the research team has the opportunity to identify overall themes and discuss specific surprising events that occurred during the observation/interview. This is a vital step in developing insights from observation as the team can review what was seen and heard in detail. A list of key takeaways, surprises, and/or behaviors should be listed. These sessions can assist directly with the coding of data for data analysis and can provide a quick summary of each visit. The debrief notes from each observation can serve as an initial guide for data analysis.

Discussion Topics include:

- Main themes for this observation/participant, *"I noticed they always..."*
- Ways in which this observation was the same or different than others. *"They did XXX just like BBBBB."*
- Surprises, *"I had no idea, GGGG happened."*
- Ideas or specific insights *"Eureka..."*

4.6 DEVELOPING A DATA MANAGEMENT PLAN

To avoid having lost data, before entering the field, it is critical that the team makes decisions about how they plan to manage their data. The importance of data management throughout the course of a study will ensure that the team is organized and that the team is prepared for data analysis.

After an observation or interview in the field data will come back to the team in various formats, such as field notes, audio, photography, and/or video, documents or artifacts. Each of these can easily got lost or damaged. This is discussed in detail for data analysis.

Managing Field Notes

- Use a dedicated field notebook and record notes as legibly as possible, so that other team members can read it.
- Scan or transcribe field notes after the observation or interview and share across the team.

Managing Audio, Photography, and/or Video Files

- Use designated devices for gathering data—refrain from using personal devices for data collection.
- Determine where your data are to be stored during the course of the study. It is suggested that you store your data on an online server. Keeping a backup on an external hard drive may also be helpful.
- Get into the habit of uploading data collected directly after an observation or interview. Losing data occurs often and will set back a projects timeline significantly.
- Make sure that the time stamp on the device is setup correctly.
- Determine a filing structure and naming convention that works best for the team, that is, participant identification_location_projectname_date.avi

Managing Documents or Artifacts

- All documents or artifacts that are relevant to the project should be accounted for and stored properly.
- For paper documents, scan the document and share across the team.
- For physical artifacts, it is recommended to photograph the objects and store them in a location, that is, accessible by relevant team members.

4.7 FIELD ETIQUETTE

Depending on the focus of the CI study, a research team may find themselves in varying locations of care: operating room, intensive care unit, emergency department, or a patient's home. Each location has a different culture, different dress requirements, and a different pace for completing tasks. The etiquette for each location is described below with contributions from fellow authors with experience in conducting CI studies for medical device development.

Prior to joining in observing care, the research team should ask for a summary of the patient, what they are about to see, and perhaps the conditions that brought the patient to the requirement of this procedure (Figure 4.12). Taking 5 minutes to gain an appreciation of clinical decisionmaking and/or device selection is always a helpful background in a CI medical device study.

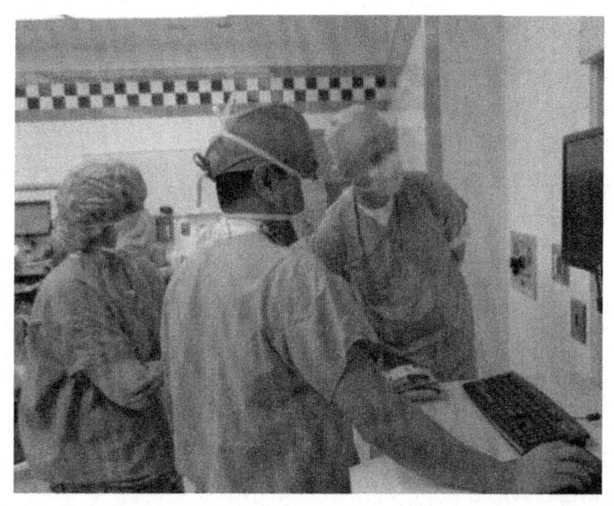

Figure 4.12 Ask the physician or nurse to review what to expect before a case or stepping into a patient's room.

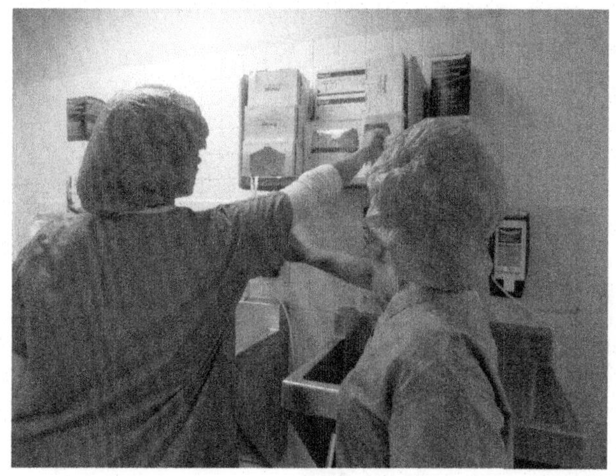

Figure 4.13 In the OR, hair*, and shoe covers plus mask are required. *Ladies with long hair might want two bonnets.

In addition, as a research team moves about the hospital there are supplies located strategically to maintain cleanliness and sterility. For example, outside of every hospital room or just inside the door, there is hand sanitizer, and just outside the sterile corridor of the OR, there are masks, hats, and shoe covers (Figure 4.13).

4.7.1 OR Etiquette by Beth Loring

The OR is a place where life and death decisions are routinely made and dedicated people are working together—sometimes at high speed—and according to specific protocols. Being invited to observe is a privilege, so you should know the proper etiquette ahead of time to avoid embarrassing or potentially hazardous mistakes.

OR Attire

First, you will need to change into the appropriate attire. This includes a scrub top, scrub pants, hair cover, mask (or face shield), and shoe covers (Figure 4.14).

The mask should be form-fitting to your face with both sets of strings tied (The part of the mask with a stiff bendable edge or metal strip is the top; take the metal strip and place it on your nose and press down, then tie the ties.). Make sure that your head and facial hair are completely covered (if you have long hair, a ponytail helps).

Avoid excessive jewelry. If you wear earrings, these need to be covered as well. Don a new mask for every case; do not walk around the hospital with a mask around your neck.

Operating rooms are intentionally cold to help keep supplies sterile and the OR staff comfortable, so you may want to wear a tee shirt

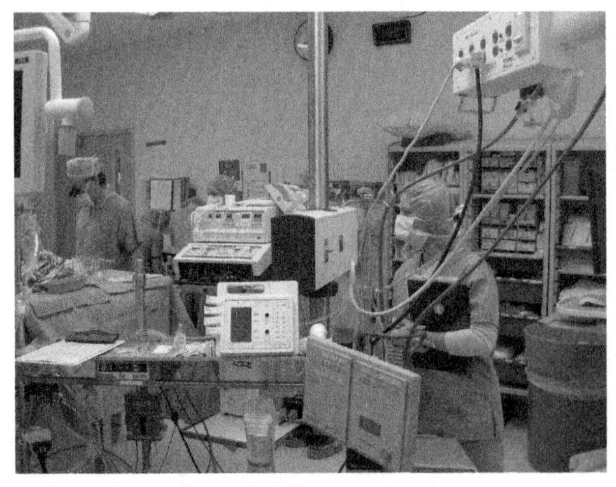

Figure 4.14 Beth Loring watches surgical preparations.

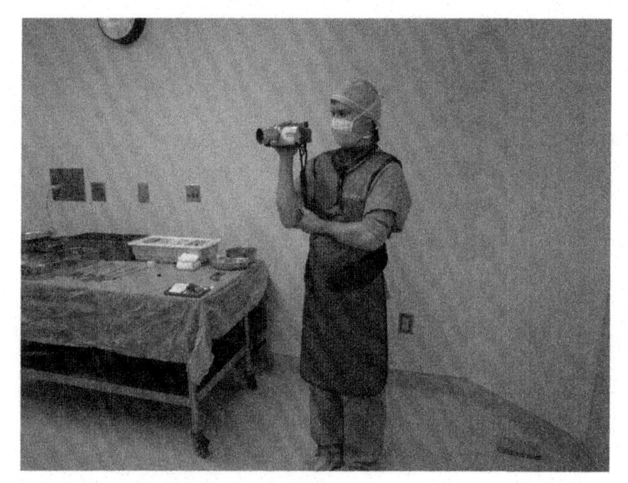

Figure 4.15 Observer wearing protective gear.

underneath the scrub top. If you are observing in a room that has the laminar flow air handling system on (e.g., total joint replacements and spinal surgeries), it can get particularly cold.

Be sure to wear flat, comfortable shoes and socks: you will be standing for a long time on a hard floor. Note that in many countries outside the USA, you will be provided with shoes or sandals to wear.

If the procedure involves radiography (X-ray), be sure to wear a lead apron. You may also choose to wear a lead thyroid shield to cover your neck. The OR staff will show you where the guest aprons are located. If lead is not available for you for that case, either step out of the room during imaging or stand behind a non-sterile person who is wearing lead. Out of necessity, lead aprons are heavy, which can cause some lower back pain if you are susceptible. On the plus side, they help to keep you warm (Figure 4.15).

Be aware that most hospitals have "red line" areas outside of the actual OR where scrubs, shoe covers, hair covers, and masks are required. They may also have a rule against wearing scrubs outside of patient care areas unless covered by either an approved jacket, or white coat.

4.7.2 Before the Case

Sometimes the number of observers allowed in the OR is limited, and this includes the sales reps who need to be there during the case. Be sure to ask how many people are allowed in the room at any given time. In some cases, I have had to take turns with my research colleagues in the

OR. Sometimes, the person waiting outside can observe through a window in one of the doorways, but be very careful not to get in the way of people going in and out.

Ask what you are allowed to bring into the OR, for example, your camera bag and clipboard may be fine, but you may be asked to leave backpacks and briefcases outside. This can be a concern if you carry valuables. It is better to leave your valuables in the car and carry only what will fit in your scrub pockets or camera case. If you are asked to leave items outside the OR at the last minute, either ask where to put them or find a spot where they will not be in the way.

Unless you have special permission, wait outside the OR until the patient is prepped and draped. In some cases, you may be allowed to observe the setup, be asked to leave while the patient is prepped and draped, and then be invited back in. Always ask permission before entering the OR.

4.7.3 In the OR

Learn the roles of the different personnel in the OR: surgeon, physician's assistant, resident, scrub nurse, circulator, anesthesiologist, and so on.

Act respectfully toward every member of the surgical team, regardless of his or her role (Figure 4.16).

Once in the OR, introduce yourself to the circulator and explain why you are there or, if he or she is busy, wait until you are asked. Provide

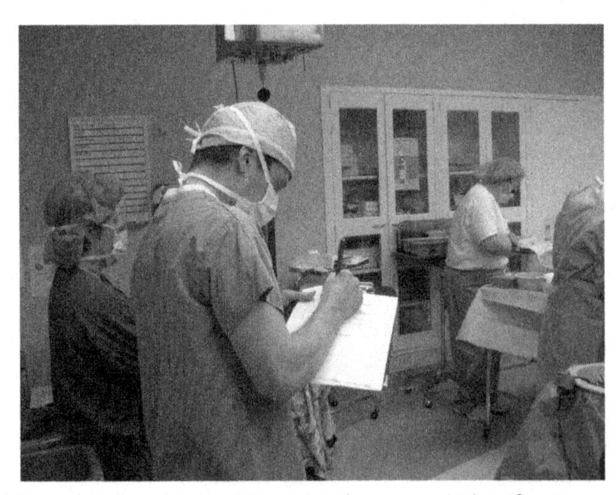

Figure 4.16 Researcher making quick notes of personnel identification.

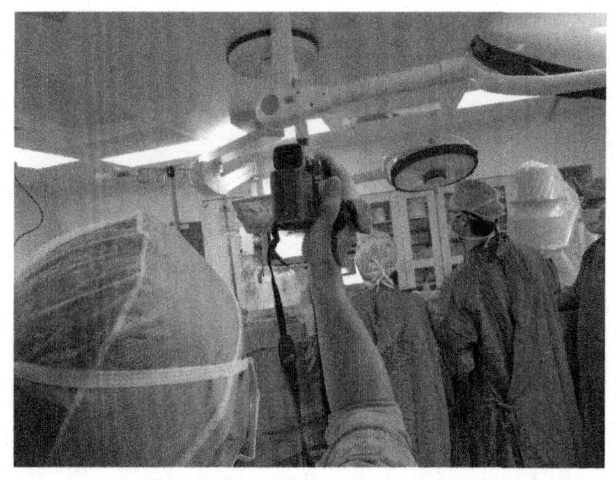

Figure 4.17 In this view, the researcher is located against the wall, as a result, they will need to elevate the camera in order to view the surgical site. This is a good perspective for recording clinical team interactions.

your full name or a business card, so the circulator can document who is in the room for the case. You may also be asked to write your name on the white board.

Often the staff will not know where you should stand until the patient is draped and the equipment is arranged. The safest bet is to stand against a wall out of the way and wait. Chances are the circulator will point out a good place to stand, or you can ask. It can be tricky to find a place, that is, out of the way, but still allows a good view and camera angle, but always defer to the needs of the surgical team (Figure 4.17).

If you have a briefcase or camera bag, tuck it out of the way in a corner or behind you near a wall. Just do not trip over it.

Stay Away from the Sterile Field

The sterile field is defined as the area three feet in all directions from the draping on the surgical table and two feet from a gowned OR team member. It includes the back table with the instruments, the mayo tray, and anything else that is blue or draped in clear plastic (such as the X-ray C-arm). Stand at least three or four feet away from the sterile field. Never get between the sterile field and sterile instruments or the gowned team.

Keep your hands close to your body and do not touch anything. Never reach, lean, or hold your camera over the back table or anything else sterile because skin can shed from your hands or arms onto the sterile instruments and cause contamination. If you are unsure whether or not something is sterile, assume it is. Likewise, assume that anything lying on the floor is contaminated and do not touch it.

A good rule of thumb is, if you do not know, ask!

While Observing

If you need to move around to get a better view, do so very slowly and carefully. There are power cords and other tripping hazards in the OR, and you do not want to accidentally bump into anything or anyone. Even while viewing the action through a video camera, stay aware of your surroundings and be careful (Figure 4.18).

If you are new to observing surgeries, certain parts of the procedure may bother you. For me, it was the initial incision, so I just looked away at that point. There can be unpleasant smells as well (e.g., from electro-cautery); some people put Vicks VapoRub™ under their nose or take a cough drop to mask it. Fortunately, the more you are in the OR, the less it bothers you; you become accustomed to seeing the insides of the human body, including blood, bone, and tissue. Keep in mind that the surgeon is in charge and the patient is well cared for. If you feel weak or light-headed, sit down or step out of the room (Figure 4.19).

Many surgeries, especially laparoscopic procedures, require a darkened room so the surgeon can see the video monitor. Avoid using flash photography because it can distract the team.

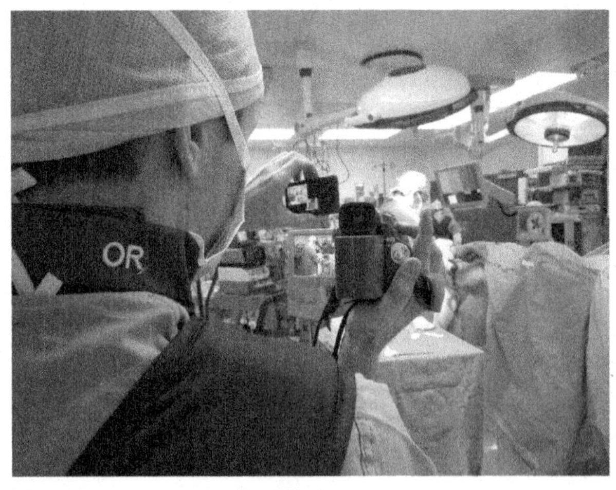

Figure 4.18 Managing cameras in the OR.

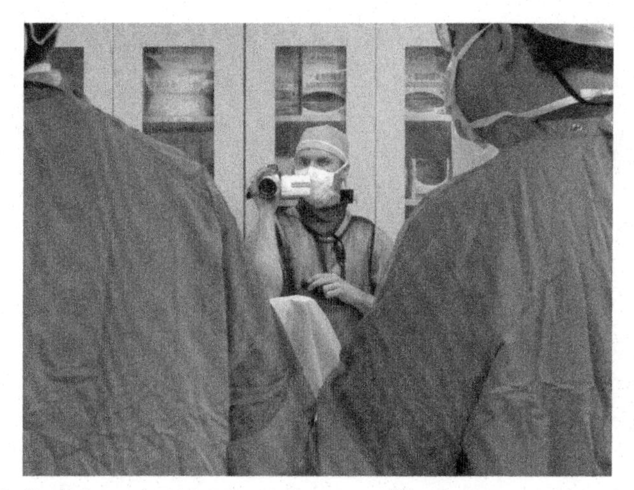

Figure 4.19 Find an unobtrusive spot. Along a wall is always a good option.

Inquire whether it is okay to ask the surgeon questions during the procedure: do not just assume it is allowed. The OR staff will typically not talk to you during a case because they are trying to listen to what the surgeon is saying, and it is hard to hear the doctor if they are talking and listening to someone else.

During many operations, there is a critical step when the team needs to concentrate and call upon their highest level skills. Try to avoid talking, asking questions, or moving during these times. In general, keep conversation to a minimum to avoid distracting the team, and use a low voice when speaking to colleagues.

If you see that something that makes you uncomfortable, such as a potential adverse event, stop video recording and follow the directions of the surgical staff. If you are ever asked to leave the OR for any reason, do so immediately and take your things.

When the case is over, thank the surgeon and staff, and make sure you have all your belongings. If you are planning to interview the surgeon afterward, ask where he or she would like you to wait.

Other Practical Advice

A day in the hospital can be very long: 12 hours or more if you observe multiple cases. You may not get a chance to have a meal, so always eat breakfast and consider bringing snacks to be eaten outside the OR (e.g., in between cases in the staff lounge).

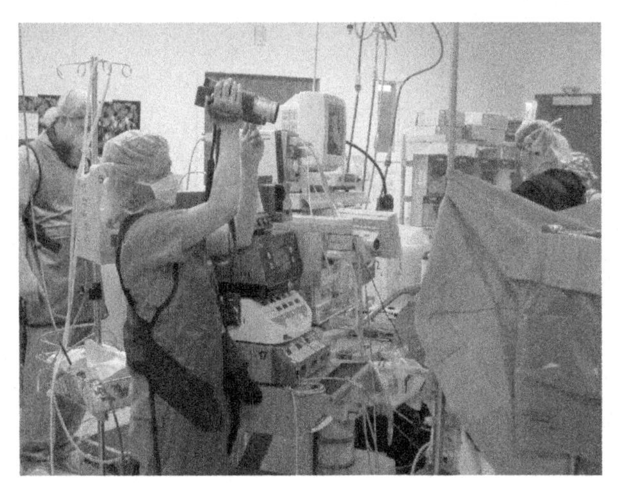

Figure 4.20 Days are long and can get exhausting.

Expect to wait around a lot. Sometimes surgeries do not start at their appointed time. Ask where you should wait, and be patient. While you are waiting, stay out of the way of the busy staff and maintain a professional demeanor. Respect the patient's confidentiality and observe the "red line" areas at all times (Figure 4.20).

When observing surgeries in other countries, be sure to orient yourself to the customs, culture, and conventions of that location. These may be considerably different from those you are familiar with.

In summary, when conducting research in the OR, always be respectful, professional, and flexible. Follow the rules and you will gather great data. Good luck!

4.7.4 General Hospital, ICU, Emergency Room Catheterization Laboratory, and Etiquette

The general hospital area is open to the public and can have a patient's visitors milling about (Figure 4.21). Likewise, while the ICU (Figure 4.22) is more controlled it too can have patients families nearby and emergency departments, especially level 1 trauma centers (Figure 4.23) are almost always very busy and filled with people.

The etiquette for conducting research in these areas is similar. They all require the same patient consent, expect research professionals to dress professionally that may or may not be in scrubs. For example, in the ICU scrubs are warranted when they are performing a sterile procedure bedside, such as placing a central line; however, for general observations,

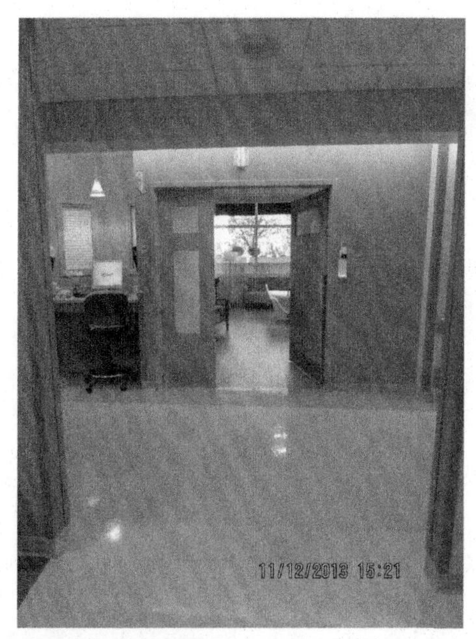

Figure 4.21 General hospital room areas.

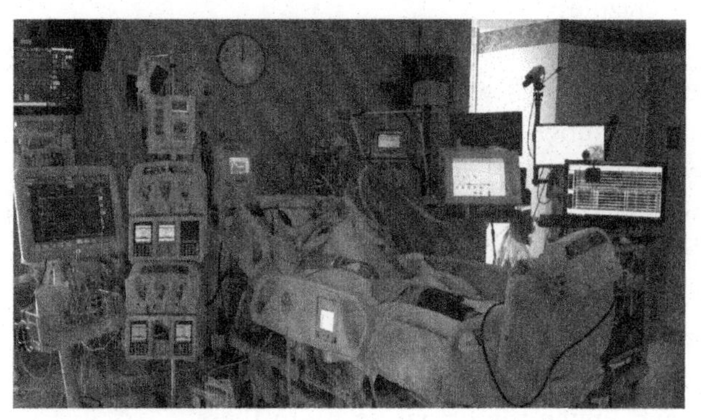

Figure 4.22 Intensive Care Unit.

such as the use of monitoring equipment, it may not be necessary to dress in scrubs. In the emergency department, scrubs are almost always worn in preparation of an unforeseen episode by a patient or trauma incident.

These environments also require the research team bring as little with them as possible and be as unobtrusive as possible. Behaviors that should be avoided include:

- Adjusting any hospital piece of equipment or device at any time, even if a clinical provider asks a member of the research team to do it

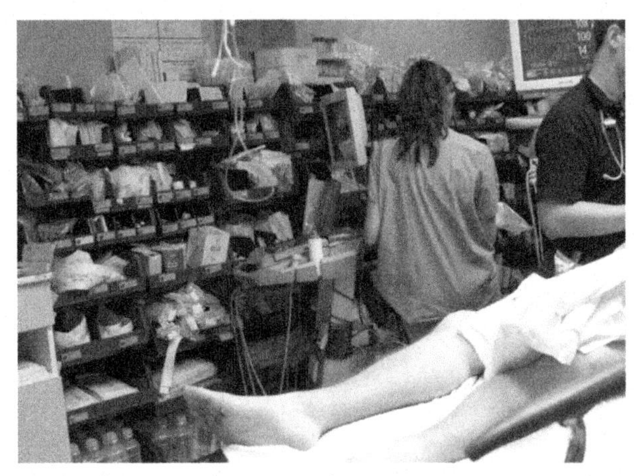

Figure 4.23 Trauma bay in an emergency department.

- Assisting the patient in any form, unless it is to protect them, such as protection from a fall incident related to bad egress
- Avoid gawking: this is very difficult in the emergency department. There is much to see, but likely none of it has relation to the research. There is always drama around the corner in this location of care, and it is easy to get wrapped up in the craziness.

Upon arrival to these locations, the research team can expect to initially find the charge nurse and notify them of who they are and why they are there. The research team should show their credentials and may be requested to show the study protocol and verify permission to observe. The charge nurse or escorting physician can be helpful in gaining patient consent for photography from either the patient themselves, or the patient's family. During procedures, the research team may want to ask where to stand, as every room is laid out differently and can get packed with patient belongings. Lastly, because these rooms can be small it can be difficult to see what is going on and unlike the operating room there really is not a good opportunity to set up an overhead video camera. It is perfectly acceptable to ask permission to lean in taking care not to violate the sterile field (any draped area) or touch any equipment.

The catheterization laboratory or interventional radiology suite is used for cardiology, neurovascular, and peripheral vascular procedures. Regardless of specialty, the patient therapy space will be a dedicated larger room with fluoroscopy equipment, including a dedicated patient table, a bank of monitors, sterile table for equipment, and occasionally space for anesthesiology to attend more complex cases (Figure 4.24).

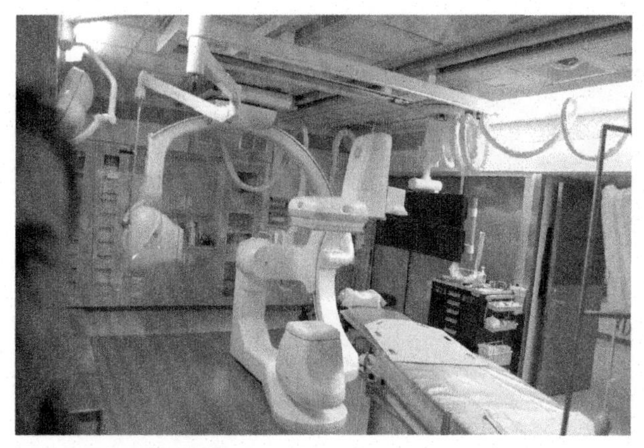

Figure 4.24 Interventional radiology suite patient care area.

Figure 4.25 Interventional radiology control room view of patient care area.

In addition, there is a control room. When arriving to observe in the catheterization laboratory, most likely you will be requested to check in with the control room. The control room is the hub of the procedure (Figure 4.25).

It is where the physicians prepare for the procedure, have directed conversations with supporting nurses and technicians. In addition, there are many supplies located within this room (Figure 4.26).

While observing in this space, it is imperative that the research team only brings in what is absolutely necessary. The spaces tend to be small and the control room can be cluttered or seemingly cluttered. The research team should never assume that they have access to the desk space

Figure 4.26 Approach planning for a complex case by interventional neuro-radiologists.

available in the control room, but certainly should feel comfortable to request some if absolutely necessary. The appropriate dress for this environment depends on the permission level. If planning to attend in the patient area, then scrubs are appropriate. If remaining in the control room, professional dress is acceptable.

Procedures can take an average minimum of about an hour to several hours overall. In addition, these suites are used in emergent situations, such as stroke for the removal of an embolization and in cardiac cases to restore blood flow to cardiac arteries during cardiac arrest. As such, their schedules are subject to rapid change and waiting will happen.

4.7.5 Notes from in-home CI Studies by Jim Rudolph

You have just completed your first comprehensive research plan; your objectives are clear, your methodology sound, your interview guide has been edited, and proofed for the 100 times, the recruiting is complete, and the final protocol has been signed off by all stakeholders. Perhaps you have even conducted a pilot test. You are ready to head into the field to conduct research in the home. Before you book your first flight, however, consider these practical lessons from the field, many of which I have learned the hard way:

- Plan to arrive early and stay longer than you anticipated. Many participants, especially the elderly, will want to share their life stories with you. It is really important that you take the time to listen, respond appropriately, and gently guide the discussion back to your research effort. Practice being patient, humble, and curious—you may learn more than you originally expected. If you have planned to conduct multiple observations in one day, make sure you leave enough time to travel between locations.

- Call your participant(s) before arriving to confirm that you have the correct location, date, and time. If you are using an independent recruiter, make sure you request the participant's direct contact information, as miscommunication can and will happen. Calling ahead of time also gives you the opportunity to reaffirm the objectives of your study, and answer any last minute questions your participants may have. I once showed up to an elderly couple's home 3 days too early, much to their dissatisfaction. Fortunately, the welcoming couple was very flexible and offered us snacks while they cleaned the house for the interview! Save yourself the embarrassment and discomfort—call your participants ahead of time.
- If you are bringing prototypes into the home, make sure they are durable. Research participants will interact with prototypes in ways you never imagined, so it is important they can withstand a few bumps and bruises. I also try to bring extra super glue and double stick tape for when the unforeseen ultimately happens (which it will).
- Also, if you are planning to bring prototypes into the home, consider bringing a small portable table with you to display the prototypes in front of your participants. This will save you the hassle of having to rearrange your participant's living room when it comes time to walk through your concepts.
- Bring backup of everything, including scripts, models, cameras, batteries, voice recorders, etc. The worst thing that could happen is you show up to someone's home unprepared to capture the information you set out to capture. It not only wastes your time on site, but also undermines all the work that went into recruiting, scheduling, and planning the visit. It is also a nice gesture to offer a copy of the script to your research participants and/or client during the observation.
- It is easy to stand out in public when lugging around disparate pieces of camera equipment, so make yourself comfortable by investing in a quality research bag that accommodates all your data collection materials (such as video cameras, tri-pods, and scripts) while also meeting your aesthetic sensibilities. This may sound like a rather trivial piece of advice, but you will be more comfortable and confident when meeting up with complete strangers in their home.

- Be prepared to repeat or rephrase your questions as many times as necessary during post-observation interviews. Participants, especially those with hearing loss, may not hear or may misunderstand your original question. It is important to not simply pass over a question for the sake of time. Remember, you have spent a lot of time and money planning your research effort, so it is important for you to get all the data points you need. Be patient and ask or rephrase a question as many times as you need to foster comprehension and gather an appropriate response. If you sense the participant is getting frustrated, move on to another topic and come back to the difficult question later in the interview.
- Expect last minute cancellations and, just as importantly, have a backup plan. No matter how much time and effort you put into your recruiting and planning phase, people will cancel just as you arrive to their doorstep. When this happens, consider a few alternatives: (i) Call the recruiter to see if any other participants live in the area (maybe they were screened because their particular profile had already met quota); (ii) Use the time to review participant responses to date; and (iii) Brainstorm solutions to some of the challenges that have been identified during previous CI observations.
- Expect the unexpected and be flexible. Unforeseen problems are going to happen—models will break, people will cancel appointments, and participants will deal with health problems while you are visiting. Be prepared and have a plan for what should be given priority. Maintaining your research participant's health, safety, and privacy is your top priority. Gathering data for your research effort is a close second.

4.7.6 International Contextual Inquiry Travelogue by Sean Hagen

There are some good practices to apply when planning an international study because there are typically more moving parts and commercial risk. The following travelogue stories may not be "good practices" exactly, but they are experiences that may be insightful for planning studies abroad. These are some lessons learned based on conducting contextual inquiry (CI) studies in 18 countries, including clinical institutions as well as home healthcare scenarios. The anthropological generalizations are anecdotal and relative to a US baseline (Figure 4.27).

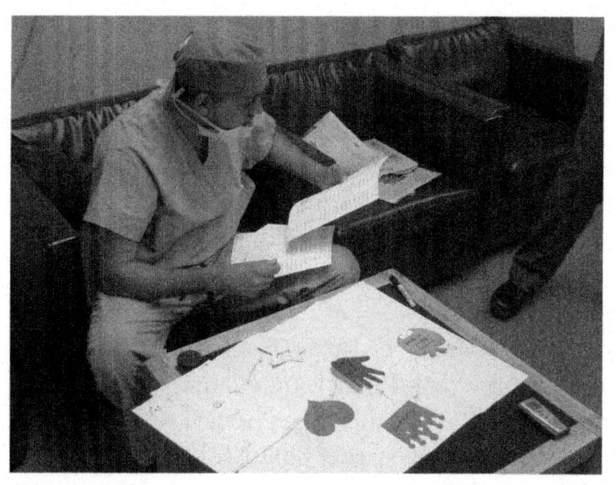

Figure 4.27 A projective mapping interview exercise utilizing metaphoric images is language independent, but requires a clear translation of the user's intended meaning for the images—Brussels, Belgium.

Research the Culture First

There are subtleties in every culture that can make a big impact in your ability to integrate into a contextual environment, especially if the goal is to blend in for an ethnography study. Here are some example stories from our travels of how important it is to be especially prepared with background research on the targeted culture:

In most of Europe, a doctor who is a professor should be addressed as "Professor," not a "Doctor." The title of professor is more prestigious than doctor there. Also, the relationship between the sales rep and the doctor is far more personal in many European countries and observing unbiased behavior is unlikely when the rep is present, more so than in the USA.

We observed several gastric bypass and gastric banding procedures over a week at four cities across France. Often the rep would be in the room during the first half of each day to introduce us, then leave to go about his business elsewhere. A common occurrence would be to see a change of instrument brands (in this case a circular stapler), when the rep was gone.

The sales rep relationship with the physician in Japan is also very different than in the USA because the social status of a physician there is extremely elite. We were running a study on transradial access for interventional cardiology and were surprised to see the rep following the doctor with a dish to catch his cigarette ashes. Then, against my coaching, a member of our team and the client's product manager were observing in the catheterization laboratory, which requires wearing lead garments—not good to try that all day in dress shoes (will write about proper travel shoes later). During a period of time where the physician was performing a procedure that was not applicable to the study, the team sat down. Three months later, I received a letter from the chief of medicine at this particular hospital (my counterpart at the client corporation also got the same letter). The letter stated that our team did not take their job seriously and behaved like they were on "holiday." Perception is reality.

Setting interviews times back to back in Germany or Japan is fine, but not a good idea in Brazil or Mexico—we have found that Latin cultures are often relaxed when it comes to punctuality. For example, one particular formative usability study in Brazil was supposed to start at 8:00 AM; at 8:30 AM, the surgeon and rep show up, to insist we go with them to see Iguassu falls via a short hike through the rainforest first—we had to cancel the afternoon session, but I will never forget the incredible waterfalls and always travel with hiking shoes now.

In Japan, it is typical business etiquette to have a drink with your colleagues; in fact it is expected to have more than one drink. In many cases, your behavior while intoxicated is considered a demonstration of your honest self—even if you manage to avoid the Karaoke performance. In hindsight, explaining the English interpretation of Kinki University to my counterparts in Osaka should have been avoided for the sake of not embarrassing the nurses on the following day.

At the other end of the spectrum, in India, going out for a drink to get to know your colleagues is not going to be as important or embraced as it is in Japan—be aware and respectful of religions that prohibit alcohol (Figure 4.28).

A discussion guide probing use error or past employment experience in India and Japan needs to have carefully worded semantics, preferably by a native speaker, as to not be insulting. Our client had a question in a recruitment screener that asked if the clinician had ever been terminated from a previous clinical employer; it was intended to filter out poor performers and currently unemployed clinicians. This was found to be highly insulting in Chennai, and no recruitment was successful until we removed that exclusion criteria. The conceptual difference between *use* error and *user* error has not really disseminated through corporate America and is still very new abroad as well. Many cultures are even more sensitive to fault in the business context than the USA, even though they do not have the med-legal pressures.

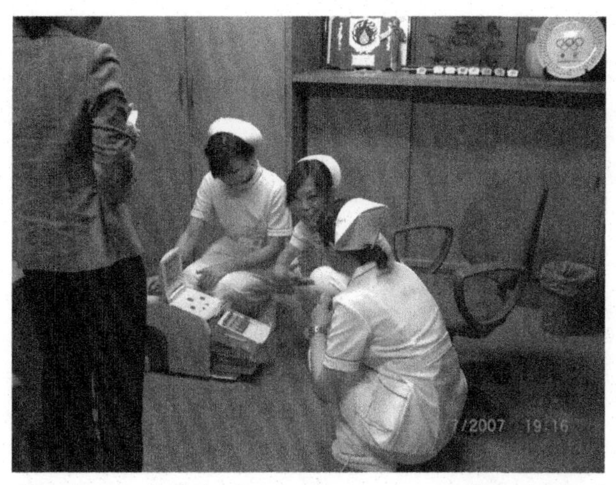

Figure 4.28 Since nearly half the homes in Japan are still tatami style, this study model had to be evaluated on the floor as well as a table—Osaka, Japan.

Lastly, familiarize yourself with a map of the places you are going to ahead of time (good to do on your flight there), and if you are bouncing from country to country, make note of the time zones (if you have phone service it should do that automatically). After working my way from the south of France through Europe and ending up in Sweden, over a two-week period, I had lost track of what time zone I was in, or what latitude. Evidently, just because the sun is up does not mean restaurants are open (it was nearly 11:00 PM), and not worrying about closing the blinds because you have planned on getting up early does not apply there during that time of year either.

Engage a Local Liaison

It is critical to have a local liaison for managing logistics and politics even if you speak the language. For example, utilizing a local liaison or counterpart to provide insight as to how to disclose and document your prototypes or study models can make the difference between your pelican cases spending a week in impound and a successful study (Figure 4.29). One simple line on a declaration form can decided if your cases are torn apart or a steep tariff is enforced.

In the 1990s, when tower minicomputers in a black finish were the state of the art, Venezuelan customs inspectors had never seen them; had we called it a machine controller instead of a medical device, it would have saved us from having to replace the hard drive in Caracas during

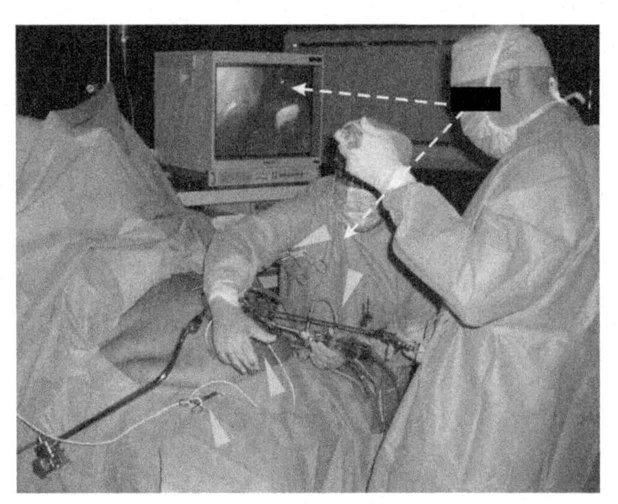

Figure 4.29 Ethnography in the OR identifying when and why a surgeon had to look at an instrument rather than the endoscopic display—Nice, France.

siesta. Our liaison had advised us to wait for him before proceeding through inspections, but we were impatient with that South American punctuality (see Research the Culture First section in this chapter).

Of course, this does not matter if the airline loses your Pelican cases. I once arrived in Amsterdam and alas, no luggage (simulated arms with vasculature, prototypes, etc.). The airline told me it would be there by noon the next day. I arrived at the hospital at 5:30 AM as planned and told the awaiting physicians the evaluations would have to be postponed to later that day. This scenario repeated every day for five days. No backup plan meant five days of sitting around doing nothing except getting caught up on emails. And no, there were no fun and games at night, since I fully expected to have to be on top of my game at 5:30 AM the next morning each day. I now try to pack prototypes into a carry-on if possible since you can replace your clothes on site.

If you are shipping or checking your study models be sure to prepare a tool kit for repairs for the very likely event something gives way to a nurse's efforts to break it the moment it is picked up. If the study model or interview exercise requires supplies, pack extra. The point is finding an Exacto knife, super glue, and Post-it notes in Hokkaido are not as easy as you might expect (although there are amazing hobby/model shops in Tokyo).

A liaison can also keep you from offending your hosts and save the day with subtle negotiations. When put on the spot to assist during a surgery in Corsica because they were shorthanded in an OR that smelled of cigarette smoke, our liaison explained, in a perfect segue, that my colleague was pregnant and could not stay to operate the camera leaving me to do that instead of being available to hold the liver retractor (she really was pregnant, just not showing).

Strategic Packing

Wearing supportive shoes is wise when traveling in general; because the rest of the world walks significantly more than what is typical in the States, unless you live in a big city like New York. It is even more important if you plan on spending all day standing in a hospital after walking to and from your hotel, train station, and taxi stand. Pack your dress shoes for the meetings and introductions, but running shoes are the norm in the clinical context these days.

I bring a shoulder bag to the hospital with a set of scrubs (vacuum sealed in a plastic bag), running shoes (unfortunately size 13s take up most of the bag), digital notepad, backup paper pad, pens, laser pointer (so as not to point over a sterile field), phone, camera (still camera, video equipment is packed separately in bullet proof cases), hand sanitizer, honoraria checks, and chargers (with outlet adapters). I learned to bring my own scrubs because they do not have as many clinicians my size in most of the world except for Northern Europe. Your name on your scrubs also helps in breaking the ice.

Traveling light will save you the stamina you need for your study. Literally running up and down stairs, through train stations is typical for Japan and much parts or Europe. Wearing the same pants a few days in a row and washing your tee shirt in the sink every other night to make room for your study documents in your roller-board could make your day a lot less stressful during transit.

Finally, another general travel insight that can save the study budget and schedule; getting traveler's gastric distress abroad is a given, pack Loperamide and ask your physician about Ciprofloxacin prior to leaving the States. It was not a surprise the salad in Mexico City caught up with me in Guatemala the next day. Although the one hour surgery observation felt like four, I made it through but had to decline on the personal tour of the local volcano by the surgeon which was not good politics. I did not expect to experience the same in Austria and had to fly in a backup ethnographer to cover for me whilst I spent two days in the hotel room feeling miserable. Since then I increase the probiotics ahead of time and take the Cipro per doctor's orders before leaving the US.

Planning Summary

After reading some of these anecdotes, you may ask, why bother with these studies abroad and the additional complications to study planning with significant added commercial risk? Obviously, the need for advancing healthcare is a universal need and medical device manufacturers understand they have to compete in a global market. However, unlike other product industries, say consumer electronics, you cannot depend on rapid iterative design cycles to get the design right through a trial-and-error model; when a life is on the line, it has to be right the first time.

From a commercial perspective, the one-size-fits-all model does not work either because every country pays for healthcare differently, has different regulations, different infrastructure, and different expectations and needs from the users. For example, Cherry blossom pink housings and playful looking cartoon pandas narrating the help screen may very well be an appropriate user experience approach in Japan but not taken seriously in the US. Conversely, a glossy black housing with white and chrome details may be a sophisticated aesthetic in the US but offensive in Japan for a medical device design (black-and-silver envelopes are symbolic of funeral condolences).

Managing the cost of global research should be addressed in a strategically creative manner to optimize the return on investment. For example, you can combine a generative ethnographic study with an evaluative user preference study as long as you complete the generative first. Informing the design and development process with explicit insights that have been synthesized from CI at the point of practice enables development to effectively assess the design requirements that are in common across global markets and those that are unique to a region or culture. Show me, do not tell me (Figure 4.30).

Figure 4.30 Observing a patient's current use scenario before introducing a user preference exercise—Sapporo, Japan.

4.8 BEST PRACTICES

- Plan on arriving 15–30 minutes before a scheduled observation. Clinical sites are often a maze and difficult to navigate.
- Ask if there is a dedicated place for belongings. Only bring what you absolutely need.
- No one will wait for researchers; we cannot interrupt patient care.
- If you feel faint or ill, tell someone sooner rather than later.
- Passing out happens. No one expects it and everyone is embarrassed by it. It is OK. You might end up being admitted to the emergency department.
- Cases may get cancelled in last minute, be patient.
- Cases can be mis-communicated. This is rare, but does happen. Try to reschedule an appropriate case to observe.
- Keep a dedicated bag with all the recording supplies.
- Have each person be responsible for maintaining their own data: collecting, transcribing, and coding.
- Document all the visits in a standard manner, and possibly use standard QMS procedures for maintaining the documents for inclusion of the design history file.
- Dress accordingly with comfortable shoes.

REFERENCE

Creswell, J.W., 2014. Research Design: qualitative, quantitative, and mixed methods approaches, fourth ed. Sage Publications, Thousand Oaks, CA.

CHAPTER 5

Data Analysis

Mary Beth Privitera
University of Cincinnati and Know Why Design, LLC, Cincinnati, Ohio, USA

Contents

5.1 OVERVIEW

"We cannot solve our problems with the same thinking we used when we created them."

~Albert Einstein

Data analysis actually starts in the field when the research team selects where to aim the camera/s and decides what notes are important enough to write down. However, further analysis, post-observation, provides the ability to collate design insights across all data collected and takes a deeper look into the field experiences. The task of analyzing data and providing meaningful conclusion can seem like a daunting task due to the tremendous amount of data collected. This is especially true for large studies or studies with multiple facets. The reality of qualitative research for the purposes of design is that not all of the information collected is actually

M.B. Privitera: Contextual Inquiry for Medical Device Design.
DOI: http://dx.doi.org/10.1016/B978-0-12-801852-1.00005-8

useful. Analysis skills are necessary to discriminate the valuable information from the noise. Some of the data will be potentially collected for another use while some information may be discarded. For example, say the research team headed out to the operating room (OR) for what was to be a target procedure only to find out that the surgeon changes the devices they use mid-procedure but they have permission to observe and video. If time permits, why not observe and learn. Watching users complete tasks and understand who they are is part of the inquiry and while the data collected is not in the target, the team can gain the flow of the organization and begin the relationship with the study participant. Certainly some of the data collected might fit to be analyzed while the rest can possibly be ignored and stored for another project. At the heart, the process of data analysis includes deciding which pieces of collected information are important in regard to the goals of the study and which pieces are not.

This chapter provides a detailed process by which a research team can take collected data into consideration and organize it to maximize impact (Figure 5.1). This process begins with reviewing all of the data then organizing it into a retrievable database and adequately planning staff requirements. Next, using debrief notes and field memos, themes and codes are generated based on evidence. The chapter provides specific tools for the analysis of video and images. This enables the data to be displayed visually and condensed using the best examples of evidence. Finally a discussion on tenacity and ethics are reviewed.

5.1.1 Data Management and Review

The types of data gathered in a CI study can vary widely; this includes field notes, video and audio recordings, still photography, and documents. Maintaining the origins and managing the data itself requires a system. The system can enable the research team the ability to quickly retrieve a piece of evidence, for example, photo or video of a particular event quickly. If a system is not developed and maintained, inevitably someone on the team will be sorting through all of the data in hopes of identifying something they know they have seen or heard but cannot find it.

As mentioned in earlier chapters, field notes are the combined written and/or digital artifacts used to record data during a field observation. Field notes contain information that the observer notes during an interview or observation. Field notes may be written in an interview guide

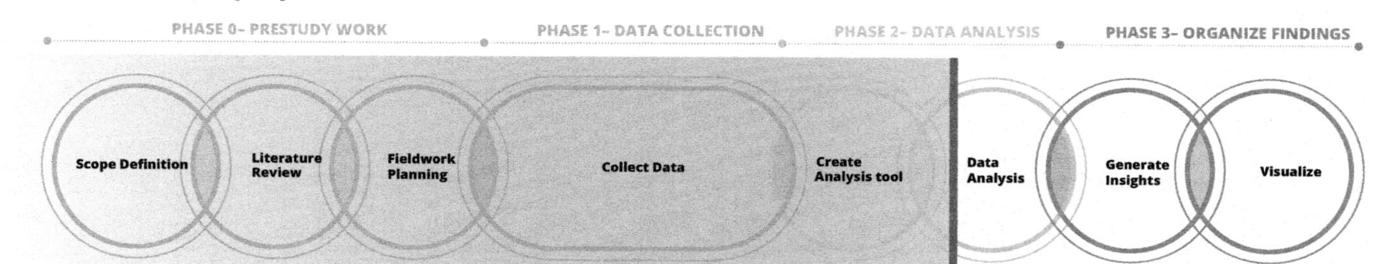

Figure 5.1 Data analysis is a detailed step in conducting a contextual inquiry (CI) study for medical device development.

and contain basic notes, diagrams, sketches, or quotes from study participants. The best habit of a researcher is to be diligent about personally managing their data. For this reason, it is extremely important to transcribe and synthesize data immediately after an observation. The information collected may need clarification or gaps in the data may need to be finalized due to limitations or constraints while in the field. It is ideal to do this the same day possibly during the debrief session, or no later than 2 days after an observation. This ensures that the information is captured with a high resolution and that it can be shared among the team immediately.

Video data will be gathered in parallel with target goals and the interview guide. A focus should be on only recording the data that will help achieve the study goals. For example, the video should record only those observations that assist in gaining a better understanding of challenges, mitigations, and opportunities for improvement, rather than any outlying side conversation or prep. This means the team must be prepared to know what they will see prior to arrival. If the team is unsure what is going on, they should ask. Immediately after an observation, the data should be downloaded from the recording device onto a server. This step is critical in order to ensure that data is not lost and to prepare the device for the next observation. Being diligent in downloading the data onto a server then wiping the video equipment clean will assure a smooth transition to the next site visit while maintaining the data.

Like the field guides and video data, it is important to also recognize still photography and documents collected along the way. Still photography should be focused on target goals and downloaded to a secure location immediately following a field visit. For documents and other items collected during the visit, before downloading, the research team may want to take the time and photograph everything. Examples of the documents or other elements collected are stickers used/not used that get placed in paper medical records or extra devices used in demonstration. For example, if comparing clinical kits for a particular procedure, the research team may collect a representative kit from each manufacturer found in the field and then closely analyze the various elements. If photographed immediately, the original condition, contents, and content organization can be preserved (Figure 5.2).

For all of information collected, standard file management and storage location is critical for easy data retrieval. A naming convention is suggested which retains some information regarding the date, location, and

Figure 5.2 Wall of devices from various kits displayed for comparison.

type of data collected is recommended. This enables quick reference for the status of data collection and overall program management. An example is presented below:

- YYYY.MM.DD* FACILITY UNIT FIELD-NOTES INITIALS.doc
- YYYY.MM.DD* FACILITY UNIT FIELD-NOTES INITIALS.pdf
- YYYY.MM.DD* FACILITY UNIT VIDEO *descriptor* INITIALS.mov
- YYYY.MM.DD* FACILITY UNIT IMAGE *descriptor* INITIALS.jpg

Once all information is located in one place, each person on the research team should begin data analysis by reviewing all of the files to understand and plan their workload accordingly. An assessment for missing information may be necessary, as during busy field visits often a download session may have been missed. The team should develop a plan and schedule for data analysis. Of note, the type of data collected will determine how the team conducts data analysis. For example, if using video as the main data capture tool, this requires time to process and code all of the video. As a rule, allow 3–4 hours of video interpretation time for every hour of video.

5.1.2 Staffing and Planning for Data Analysis

Data analysis is a tedious process and almost always the deadline for study completion looms. The team that ventured out on the field visits should be heavily involved in the data analysis as it is very difficult to discern information based on notes taken by another person or by video that has significant starts/stops. The field team can assist in filling in these gaps and providing further context. In addition, both senior and junior staff should be involved in both data collection and analysis. Avoid having junior staff collecting data and senior staff analyzing data or vice versa. Ideally, the data collection and analysis is a team effort.

The timing of data analysis is almost always slower than originally planned. As a general rule of thumb, a research team can expect to need roughly 3–5 times as much time for processing the data as the time you needed to collect it. For example, each day of fieldwork expect to spend:

- 2–3 days processing field notes/reviewing video data (writing them up, correcting, etc.),
- 1–2 days of theme generation/coding,
- 2–3 days of completing data displays and writing insights (depending on the number and type of displays).

5.1.3 Steps in Data Analysis

The following steps break down the process of data analysis. It is important to note that in CI studies, data analysis happens simultaneously with the generation of design information (insights and ideas) and data visualization (Figure 5.3).

Typical steps in data analysis include the following:

1. Generate themes and data codes.
   a. Use debrief notes: look for surprises, common elements across field team.
2. Analyze data for evidence of theme and codes.
   a. Complete a task analysis: list of tools, steps used in procedure.
   b. Select images, quotes, video snips that best represent themes and codes.
   c. Develop data displays; affinity diagram sketches or other comparative visualizations.

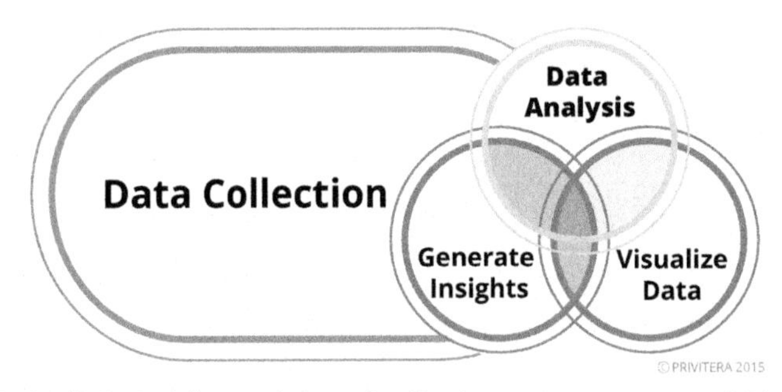

Figure 5.3 Analysis, insights, and data visualization are concurrent activities.

3. Condense themes and codes.
   a. Eliminate unsupported themes.
   b. Combine redundant themes.
   c. Synthesize data to uncover connections and causality of user behavior.
4. Review the data and select top 2–3 images, video, etc. that best represent the theme overall.
5. Write design insights or need statements (see Chapter 6 for full description).

5.1.4 Generating Themes and Coding Data

As mentioned during the debrief session descriptions, at the end of each site visit, the research team should discuss themes, the dominant elements as observed or discussed with study participants. *These themes form the backbone of data analysis. Themes can assist in the generation of codes. Codes are simply labels applied to data that assign meaning or category.* By virtue of sorting the data into themes or codes, data analysis is being conducted. In using themes and codes, the data itself is more retrievable and easier to condense bulk data into meaningful insights.

For definition:

A **theme** is a dominant behavior, idea, or trend seen throughout a site visit.

Example: Air bubbles get caught in arterial lines

A **code** is a label applied to the data under a particular theme that assign further meaning or category.

Example:

Descriptive code—air bubbles are small and difficult to see

Emotional code—"I hate dealing with bubbles, it takes too long."

Sequential code—Step 1 and 5 includes remove bubbles

Figure 5.4 shows the connection of themes and codes. The codes assist in adding further detail to the themes generated. Themes are simple statements whereas coded data may include quotes, images, and/or observations. In generating themes and codes, the entire team should be involved with the goal of codeveloping consistent definition and understanding of each theme and code. Once collated, the definitions can be maintained in a shared document for reference.

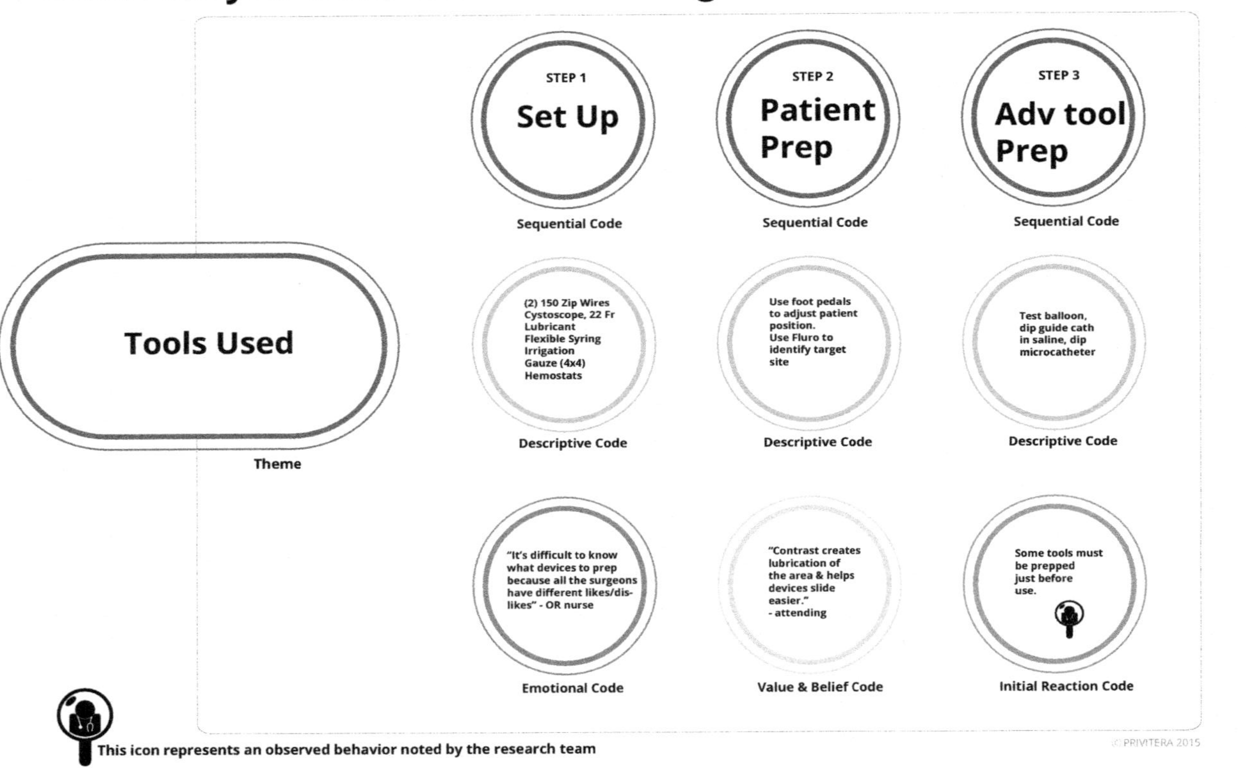

Figure 5.4 Example of coded data under a theme.

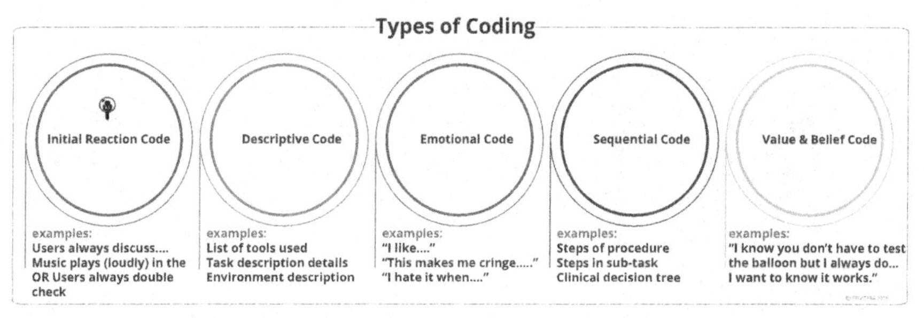

Figure 5.5 Types of coding.

5.1.5 Creating Themes and Codes

At the onset of the study in the development of the protocol, the initial set of questions and study goals can drive the development of themes or codes that are simply hypothesis driven. For example, a team may want to discover if a user truly performs a task that they have anecdotally heard was common but has not been reported by sales force or others with direct contact with medical device users. Further, the secondary research on current practices can generate codes. For example, if a CI study is based on a surgical procedure there are often surgical texts or manuals that a young surgeon may use to inform their practice. The steps of the surgical case as represented in this body of literature can serve as the initial sequence of steps. There is a caution in this approach however and that is to ensure that what is represented in literature is fully referenced and the references are maintained. If not, then the source of what was discovered prior to the study can muddle the results of the contextual study.

The alternative to generating the codes prior to the onset of the study is to let the observation and interview discussions drive all coding as a result of the data itself, thus producing a case analysis for comparison (Huberman and Miles, 2002; Yin, 2009).

This approach relies on logical groupings generated through a variety of approaches including initial reactions, descriptive coding, emotional coding, values coding, and sequential coding. Each of these is described above (Figure 5.5).

Initial Reaction Codes

As data is read or experienced, the researcher notes patterns, processes or commonalities, and relationships as a result of their own personal interpretation. This may include remarks or analytic memos that describe the

Figure 5.6 Icon indicating that finding is result of research team observation.

opinions or assertions made by the study participant. Initial coding is meant to be relatively quick glance at the data and may include or be related to other types of coding.

Figure 5.6 is a quick icon to add to initial reaction coding and indicates that the finding or data point represents notes taken by the research team. Thus, it may need further investigation to prove a valid interpretation. By clearly indicating the origin of the data, such as icon use, further investigation can be initiated and the data point traced to a research notebook.

Example: "There was confusion at the start of this case; the surgeons consulted each other in great detail on what devices would work best."—Observation team

Descriptive Codes

These are codes that detail an item for the purposes of inventory or category. These are useful when identifying themes related to tool use or nonuse. Descriptive codes can be the names of items or words used by the participants for the purposes of tool identification (Saldana, 2009). For example, a participant may "place a patient in reverse Trendelenberg" which means that the patient's table is head-side down. Having a descriptive code of Trendelenberg may generate insights regarding overall patient positioning across the data set.

Example: "Both the catheter and the users hands are wet and the device slips as they insert it."—Interventional cardiologist

Emotional Codes

Mood of the study participant and the workplace always has an effect on the efficiency and/or enjoyment of completing tasks. As such, this is an important type of code. Emotional coding is subjective coding that labels either what was inferred by the researcher or communicated by the study participant (Saldana, 2009). These are useful in developing an understanding of how a user relates to the tools or procedures emotionally. In the practice of medicine, occasionally those who practice may have egos that make it challenging to discern opinion. For example, a particular part of a procedure may seem impossible to the research team and the difficulties may be communicated directly. However, those challenges faced daily may be the exact part of the job that is loved by the practitioner. For best practices, when opinions are inferred, the origins of the inference should be noted and possibly indicate further exploration for confirmation.

Example: "It's difficult to know what devices to prep because all the surgeons have different likes/dislikes."—OR nurse

Value and Belief Codes

Values are the ideas that study participants hold as important and beliefs are concepts they hold as truth (Saldana, 2009). Identifying attributes that users communicate as their values or beliefs can further determine their attitudes and opinions regarding their behaviors and interactions with others. There are many examples throughout the history of medical device development wherein clinical provider beliefs directly influenced the devices used for the treatment of disease or ailment. Most notably, the use of bloodletting which is an operation with a 3000-year history wherein practitioners believed evil spirits caused the ailment or disease. The process of bloodletting was to facilitate release of impurities and resulted in the development of elaborate medical devices for centuries. In contemporary practice, the phrase "its always been done this way" can lead to root behaviors and opinions which may no longer be consistent with scientific discovery. However, since values and beliefs directly drive device use choice, they are important discoveries in a CI program.

Example: "Contrast creates lubrication of the area & helps devices slide easier."—Attending

Sequential Codes

Organizing information from the onset of one task to the completion of another is sequential coding (Saldana, 2009). It is useful in discovering

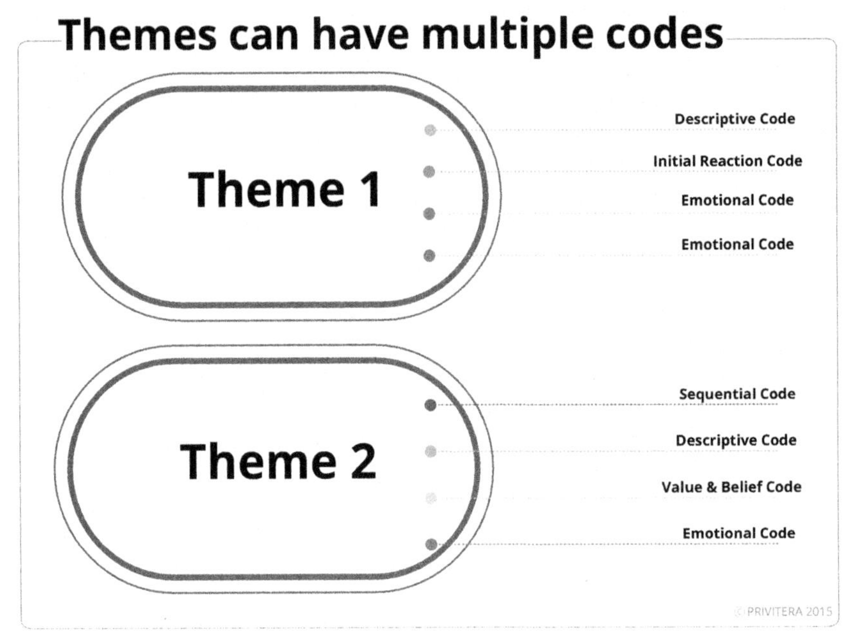

Figure 5.7 Themes and codes relationship.

the processes that involve action and focus on the dynamics of time. Throughout device use there are often other devices or people interacting and intertwining in order to facilitate a procedure or goal. Sequential codes can be used to describe the order in which they happen, emerge, change, or occur. This is the most common form of coding in CI studies.

Example: Steps of the procedure:

1. *Equipment preparation*
2. *Patient preparation and imaging*
3. *Preparation of guidewire and guide catheter*
4. *Gain access, advance to aortic arch then target artery*
5. *Prepare balloon catheter*

In the process of organizing and coding the data it is possible for multiple codes to be under the same theme.

Figure 5.7 shows two themes each with different codes associated with them. In analysis, the study objectives will drive the discovery of themes that are most relevant and the codes will drive the evidence and details of the findings. When put together the data can be condensed in order to develop design insights for each theme. This is discussed in detail within Chapter 6.

5.2 TOOLS FOR DATA ANALYSIS

Conducting data analysis can be completed by hand using readily available spreadsheet software or specific qualitative research software such as NVivo. Both have their advantages and disadvantages. Spreadsheet software is commonly used by most medical device professionals and therefore enables participation in data analysis without the need for learning special software. NVivo is extremely helpful since its origins are qualitative research and thus it is designed with features that ease data analysis. Another tool in data analysis is an affinity diagrams (Beyer and Holzblatt, 1999) wherein like items can be sorted in buckets can also be used. This is most often completed using post-it notes. For medical device development, this method is limited as it is difficult to get adequate detail while maintaining context on a post-it note. However, after some data has been coded and recorded this method can be appropriate for exploring design insight generation.

An example analysis template designed for use with spreadsheet software is seen in Figure 5.8. This spreadsheet has the details of the site visit along the top. This includes details regarding the procedure, the observation identifier, the participant, tools used, specific details regarding the patient presentation, an area dedicated to other details purposefully left blank and is intended for information such as hospital location, a space for special considerations, and a listing of all personnel involved. Across the horizontal axis is a focus on procedure steps from the onset of the procedure through closing. Below the challenges, mitigations, insights (if they occur during analysis), quotes and notes are listed. This template is intended to be generic and should be modified according to the details of the study. Additionally, this template is designed to expedite data visualization through procedure mapping which is detailed in Chapters 6 and 7.

An alternative template, Figure 5.9, demonstrates a template using specific code references. In this template, the primary details of the observation are maintained in the upper right, the sequence coding along the top axis can provide the steps of the procedure observed then listed below are the timing, value and belief codes, the emotional codes, and the descriptive codes. In this arrangement the coded information as they relate to the procedure steps can be highlighted. This template is most appropriate to use when the discovery of detailed opinion is paramount. When analysis is completed using this template, the opinion itself, the context and its relative aspects can be identified.

The team should plan the best means to collaborate for data analysis and consider developing their own specific template for a study.

Procedure	obs #	Participant	Tools	Patient presentatio	Details	Special consideratio	Personnel involved

Time						
Procedure Steps	Start > Prep		Prep > Major event		Major event > Closing	
	Time Stamp	Procedure Step	Time Stamp	Procedure Step	Time Stamp	Procedure Step
Challenges						
Mitigations						
Insights						
Quotes						
Notes						

Figure 5.8 Observation data analysis template version 1.

Site Visit Details		Sequence Code 1	Sequence Code 2	Sequence Code 3	Sequence Code 4	Sequence Code 5
Participant						
location						
Procedure						
Date						
Time	definition					
Value & Belief codes	definition					
Emotional codes	definition					
Descriptive codes	definition					

Figure 5.9 Observation data analysis template version 2.

5.2.1 Time Tracker Tool by Jim Rudolph

Time tracker (Figure 5.10) is a quick, affordable, and accessible process we developed at Farm Design to help us visualize, explore, and analyze procedural data. By turning quantitative time-based data into visual information we are able to more effectively identify patterns, trends, and

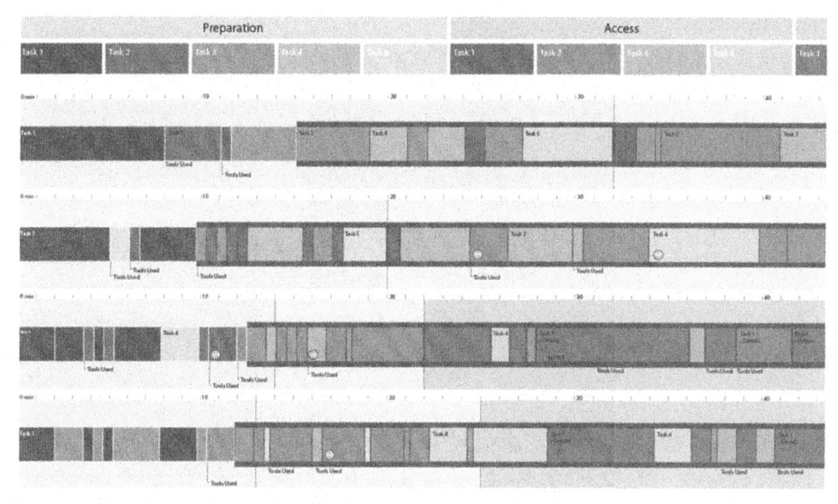

Figure 5.10 Time tracker overview.

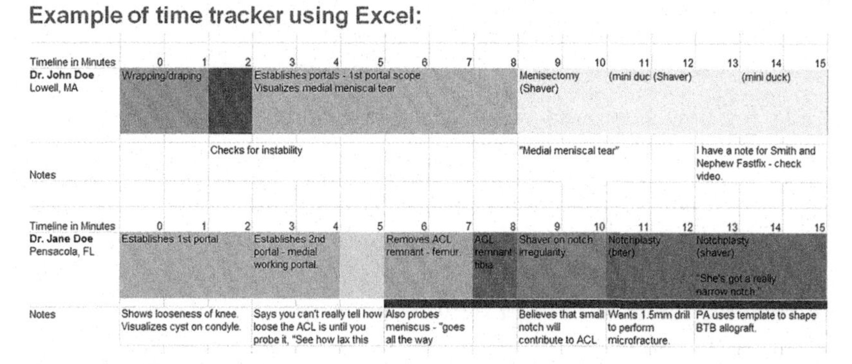

Figure 5.11 Example of time tracker using Excel.

differences between surgical procedures. Importantly, the time tracker process can be conducted using basic software tools, such as Microsoft Excel (Figure 5.11) or Adobe Illustrator (Figure 5.12). It does not require any proprietary software or specialized skills.

The time tracker tool is best used for exploring time-based information that was captured while observing a process, such as a surgical procedure. Time-based data can also be captured after an observation, by watching video, and taking detailed notes. Your notes can be taken in increments of seconds, for quick processes that are time sensitive, or increments of minutes for longer processes. I find minutes are appropriate for most surgical procedures.

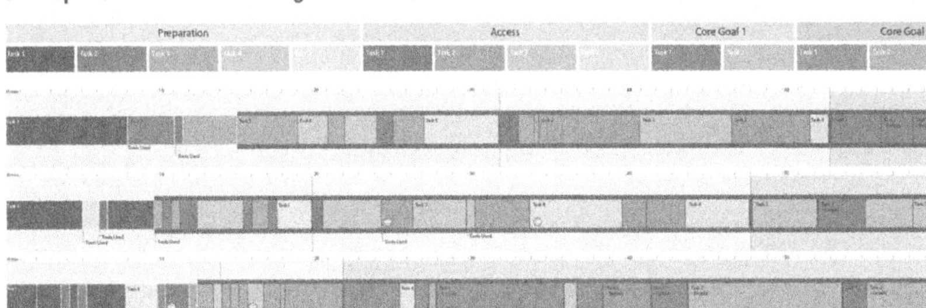

Figure 5.12 Example of time tracker in Adobe Illustrator.

Before creating your visual data, you must have detailed notes, or as we call them, time stamps. Time stamps might look something like this:

10:00–10:07—Surgeon preps patient with sterile drapes. Surgical team arranges room for carpal tunnel release (see sketch in notes).

10:07–10:10—Surgeon creates initial incision, comments "these knives are always dull" (see photo of dull knife).

10:10–10:13—Surgeon asks for new knife and then waits while the nurse goes to retrieve one.

Time stamps capture rich detail about the procedure, such as observations, quotes, usability challenges, user perceptions, and environmental conditions such as lighting and temperature.

There are a few things to consider before selecting a software program to complete a time tracker:

- No matter which program you choose, the software program must allow you to create a lot of columns (for time) and rows (for different procedures and bits of information). Importantly, the columns must be evenly spaced or the visual data will be skewed. Both Illustrator and Excel provide this capability.
- Software programs that allow layers to be turned on and off, such as Illustrator, provide an opportunity for more advanced visual search and pattern finding. By turning layers on and off, you are able to identify patterns that might otherwise be difficult to see, especially if you are capturing a lot of information.
- It is important to consider who will be using and/or contributing to the data. In my experience, most marketing, business, and engineering professionals will have Excel, but few will have Illustrator. See

below an example of a time tracker completed using both Excel and Illustrator (note: all proprietary information has been removed).

– Graphics programs, such as Illustrator, provide more flexibility (such as custom colors and icons) and control (such as positioning of different elements) for how the information is displayed, but may also require training for some individuals.

The purpose of time tracker is to:

- Explore lots of data quickly and efficiently;
- Better understand the step-by-step process for a surgical procedure;
- Compare approaches among surgeons;
- Identify steps that are taking a long time to complete;
- Identify repeated steps and back-tracking;
- Identify unmet needs, opportunities for improvement;
- Document tools used throughout the procedure;
- Document quotes and photos in a single location with context.

How time tracker works:

- Visual attributes of the time tracker (color, horizontal bars, etc.) correspond to the procedural elements we are trying to understand. See Figure 5.13.
- Color indicates the high level goals the surgeon is trying to achieve.
- Different color tones represent specific tasks the surgeon conducts to achieve his or her goals.
- Horizontal length corresponds to time on task.
- Icons are used to highlight notes of interest—for example, specific steps in the procedure.

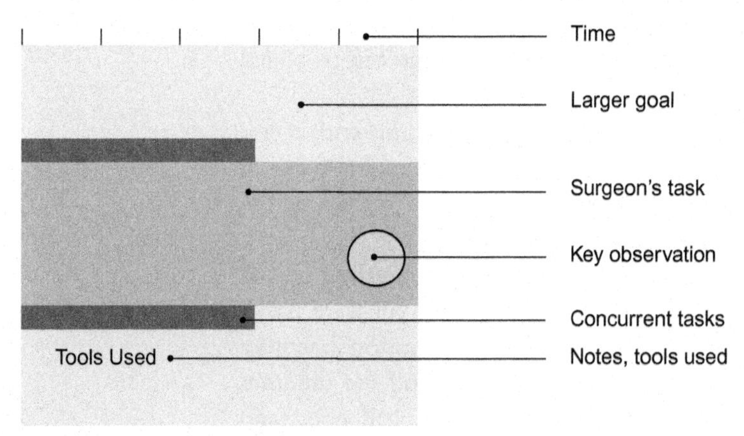

Figure 5.13 Elements in time tracker.

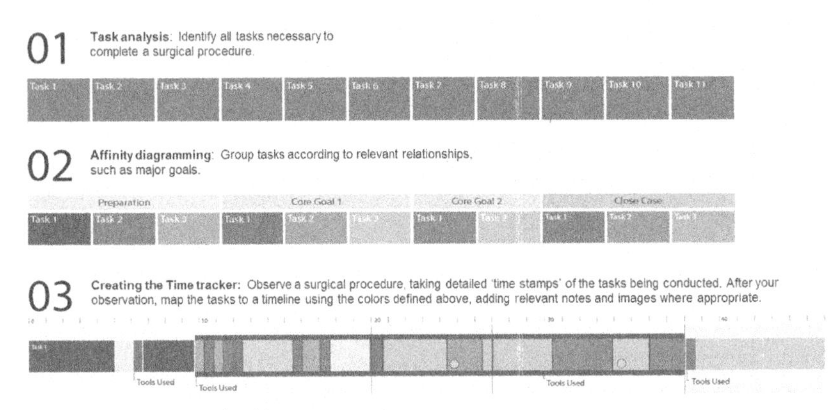

Figure 5.14 Steps to build a time tracker.

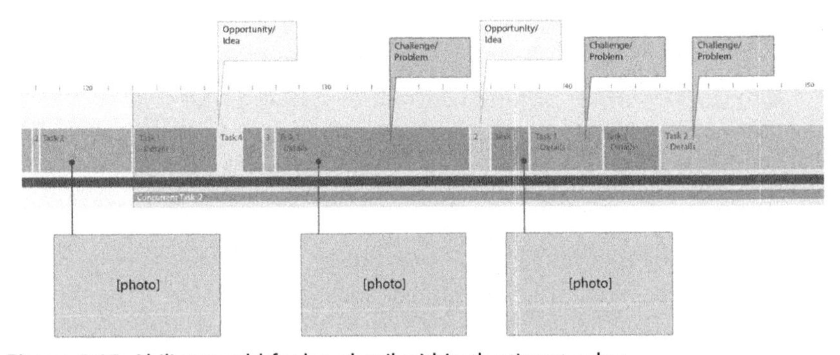

Figure 5.15 Ability to add further detail within the time tracker.

- Notes are used to capture rich details—surgeon quotes, usability challenges, product opportunities, etc.
- Additional information such as photos and video footage can be added to the time tracker for further exploration of the data.

The process of building a time tracker (Figure 5.14) begins with a task analysis, breaking down each step in the procedure. Then building an affinity diagram grouping task and finally developing a time tracker incorporating relevant observations, notes, and images.

After your initial data is mapped to the timeline, you are free to add additional information to different layers of an interactive software package such as Adobe Illustrator of Photoshop. The exploration becomes more interactive and informative if you can turn on and off the different layers. Consider adding product opportunities, usability challenges, and photos to your layers. This is illustrated in Figure 5.15.

A few additional thoughts:

- The visual time tracker is best utilized early on in the research effort.
- The data should evoke questions (why did surgeon X perform this step first?), which can be added to questionnaires for future CI efforts.
- Play with the colors to highlight specific areas or individual tasks.
- Quantify the data (time spent doing step x, number of tool changes, etc.) to validate and better support your findings.

5.2.2 Tools for Analyzing Images

In the field literally hundreds of still images can be taken which need to be sorted and organized. This process can be expedited by the creation of contact sheets. A contact sheet refers to film photography wherein all the frames of a developed roll of film would be printed in small versions for quick reference. Contact sheets can be created using Adobe Bridge or a similar program. In order to do this in Bridge, the images must be selected, then select the output menu and set desired options to export a PDF contact sheet. See Figure 5.16.

Once the contact sheet has been created the data can be coded and sorted easily. The contact sheet enables the researcher the ability to quickly scan an image and discern areas of interest. This can be marked on printed copies or on the .pdf files eventually making its way to the data analysis spreadsheet or NVivo software. Figure 5.17 includes an example of a contact sheet that has been analyzed.

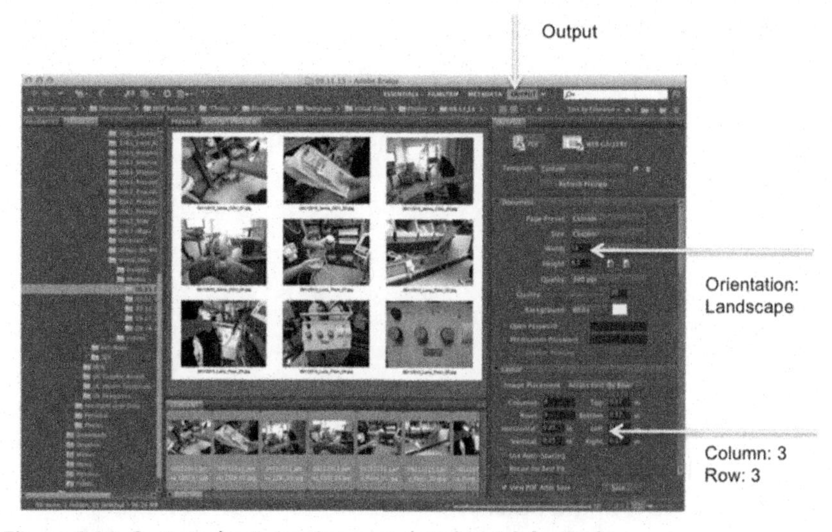

Figure 5.16 Contact sheet creation completed in Adobe Bridge.

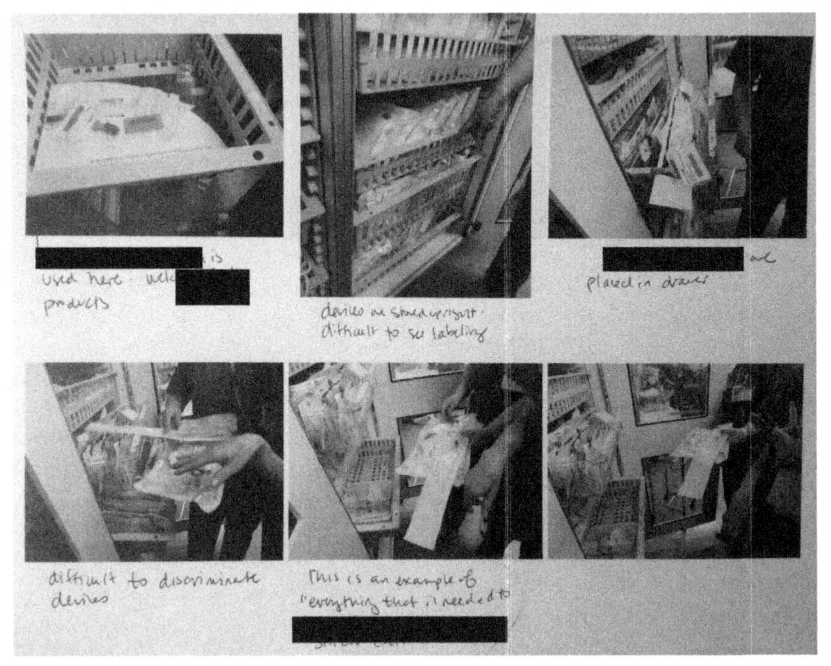

Figure 5.17 Contact sheet with observer notes ready to be coded and entered into spreadsheet.

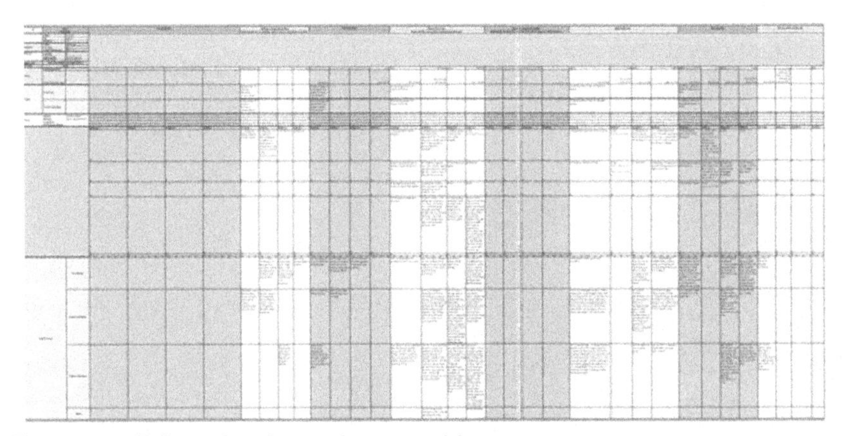

Figure 5.18 Fully analyzed procedure spreadsheet.

Figure 5.18 below illustrates a one procedure that has been coded. All of the data gathered from one procedure should be analyzed as an independent data set that can then be compared to the others.

Once completed the themes and codes can independently be analyzed across all procedures studied. As mentioned earlier, there will be instances where there are multiple examples of the same situation across the data collected. In collating the data based on themes, the best possible choice/s for communication can be determined. This is detailed in the next section.

5.3 DATA CONDENSATION

Inevitably after reviewing all of the data and organizing it based on themes and codes, the data sets within a particular theme or code may still be too large for interpretation. Taking each individual data set and condense it through a process of selecting, focusing, simplifying, abstracting, and/or transforming the data that appears in the full corpus of written-up field notes to 1–2 sentences that describe the set. In addition, within the data set of video and images there are likely several pieces of evidence that illustrates the theme. These can be organized and prioritized to include only those that represent the theme the best. Condensing the data requires an assembly from each code to be located in one place (Figure 5.19). From here, the research team can begin to search for the challenges, mitigations, causes, and opportunities for improved device design.

Once the data has been condensed, further exploration can assist in determining the following:

- Draft maps of experience can be generated.
- Interactions of all user types can be identified.
- Interrelationships of interaction points can be generated.
- Interrelationships of user experience can be generated.
- Specific recommendations for design can be generated.

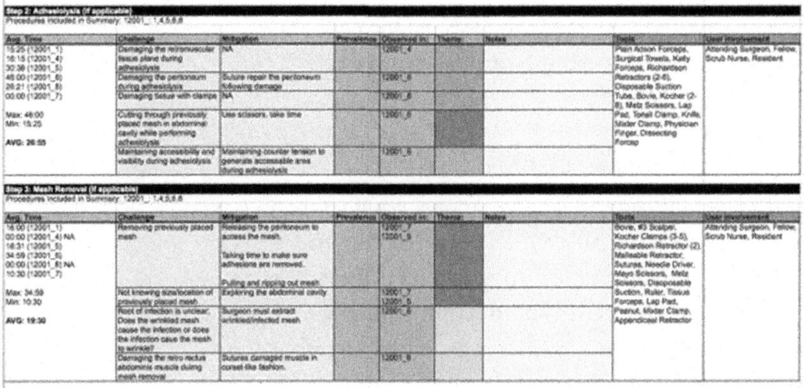

Step 2: Adhesiolysis (If applicable)
Procedures included in Summary: 12001_: 1,4,5,6,8

Avg. Time	Challenge	Mitigation	Prevalence	Observed in:	Theme:	Notes	Tools	User involvement
15:25 (12001_1) 16:15 (12001_4) 30:38 (12001_5) 46:00 (12001_6) 26:21 (12001_8) 00:00 (12001_7) Max: 46:00 Min: 15:25 AVG: 26:55	Damaging the retromuscular tissue plane during adhesiolysis	NA		12001_4			Plain Adson Forceps, Surgical Towels, Kelly Forceps, Richardson Retractors (2-6), Disposable Suction Tube, Bovie, Kocher (2-8), Metz Scissors, Lap Pad, Tonsil Clamp, Knife, Mixter Clamp, Physician Finger, Dissecting Forcep	Attending Surgeon, Fellow, Scrub Nurse, Resident
	Damaging the peritoneum during adhesiolysis	Suture repair the peritoneum following damage		12001_6				
	Damaging tissue with clamps	NA		12001_6				
	Cutting through previously placed mesh in abdominal cavity while performing adhesiolysis	Use scissors, take time		12001_6				
	Maintaining accessibility and visibility during adhesiolysis	Maintaining counter tension to generate accessable area during adhesiolysis		12001_6				

Step 3: Mesh Removal (If applicable)
Procedures included in Summary: 12001_: 1,4,5,6,8

Avg. Time	Challenge	Mitigation	Prevalence	Observed in:	Theme:	Notes	Tools	User involvement
16:00 (12001_1) 00:00 (12001_4) NA 18:31 (12001_5) 34:59 (12001_6) 00:00 (12001_8) NA 10:30 (12001_7) Max: 34:59 Min: 10:30 AVG: 19:30	Removing previously placed mesh	Releasing the peritoneum to access the mesh. Taking time to make sure adhesions are removed. Pulling and ripping out mesh.		12001_7 12001_5			Bovie, #3 Scalpel, Kocher Clamps (3-5), Richardson Retractor (2), Malleable Retractor, Sutures, Needle Driver, Mayo Scissors, Metz Scissors, Disposable Suction, Ruler, Tissue Forceps, Lap Pad, Peanut, Mixter Clamp, Appendiceal Retractor	Attending Surgeon, Fellow, Scrub Nurse, Resident
	Not knowing size/location of previously placed mesh	Exploring the abdominal cavity		12001_7 12001_5				
	Root of infection is unclear. Does the wrinkled mesh cause the infection or does the infection cause the mesh to wrinkle?	Surgeon must extract wrinkled/infected mesh		12001_6				
	Damaging the retro rectus abdominis muscle during mesh removal	Sutures damaged muscle in corset-like fashion.		12001_8				

Figure 5.19 Complete coded data grouped according to theme (by procedure step) and visualized for data condensing.

This further analysis moves the study result from that which is readily obvious to true meaning and causality.

5.4 ETHICS

With any research activity there are always ethics involved. As data analysis is conducted it is important to remember privacy, confidentiality, and anonymity are always paramount and difficult to achieve at times. For example, an incident may have happened on one particular site visit and the site may have unique identifiers that are visible in the images. Additionally, despite best practices faces and/or hospital badges may have been filmed. Care is warranted when identifying challenges of a particular user or participating site so as to avoid disclosure. There is risk associated on their behalf as these challenges are highlighted, analyzed, and commented on. Like many other issues, if taken out of the context of the study can be damaging to a relationship or career. Additionally, time has been dedicated to attaining patient consent. These consent forms should be maintained like other pieces of data and the rules identified in the document should be strictly adhered.

Lastly, data analysis is a lengthy process. There may be instances where corners get cut as analysis comes near the deadline of the overall program. The completeness to which data has been sorted and analyzed is the responsibility of the research team and should be openly communicated to the product development team. It is possible decisions related to overall product development strategy or usability objectives may be determined as a result of the study. Having slighted results may base important and potentially costly decisions on poorly analyzed data. Good data analysis maintains a record as everyone involved in data analysis is human and mistakes may happen. The team may need to simply go back and review.

5.5 BEST PRACTICES

- It is strongly advised that *data collection and analysis are activities that should be executed at the same time*. Saving all data analysis for the end is very difficult. Data analysis is actually easier to perform when the observation and interview information is fresh in your mind. Once there are more than five observations they may start to blend together and become difficult to discriminate leading to possible confusion or further need to review the data prior to coding.

- *Data analysis can be tedious and sometimes confusing.* It is confusing—there are no right/wrong only better/worse choices. Much of the data collected will be interconnected to one another. The study objectives should serve as a guide for this step in the process.
- Once the master theme and coding document has been generated, *assign an analysis lead.* This person can maintain a document of status and be responsible for updates and communication across the data analysis team.
- *Analysis should be completed as a team.*
- *Keep a list of the "working rules"* that have been determined by the team handy.

REFERENCES

Beyer, H., Holzblatt, K., 1999. Contextual Design: defining customer-centered systems. Morgan Kaufmann, USA.

Huberman, M., Miles, M., 2002. The Qualitative Researcher's Companion. Sage Publications, USA.

Saldana, J., 2009. The Coding Manual for Qualitative Researchers. Sage Publications, USA.

Yin, R.K., 2009. Case study research: design and methods. Saldana-Sage Publications, USA.

CHAPTER 6

Developing Insights

Mary Beth Privitera
University of Cincinnati and Know Why Design, LLC, Cincinnati, Ohio, USA

Contents

6.1 DEVELOPING INSIGHTS FOR DESIGN

"I never know what I do want... but I always know what I don't."
~UCMC neurologist

Interpreting contextual inquiry (CI) data (Figure 6.1) and condensing it into insights for design requires a combination of an open mindedness coupled with diligent data collection. It is a practiced skill and sometimes overlooked. An open mind is the most essential tool a researcher will use. The ability to imagine a different reality while recognizing the behaviors, actions, and thoughts of the study participants is more challenging than most individuals understand; putting aside your biases while not judging is a learned skill. Development teams notoriously enter a study with a strong perception of what they expect to see, but the point of a CI study is to use the research to inform a product development team regarding what they do not know, not what they already know. The results must first answer WHY? And then the team can collectively take on SO WHAT?

"By understanding how things work, you can start to think about improving them."
~Maurice Kanbar

M.B. Privitera: Contextual Inquiry for Medical Device Design.
DOI: http://dx.doi.org/10.1016/B978-0-12-801852-1.00006-X

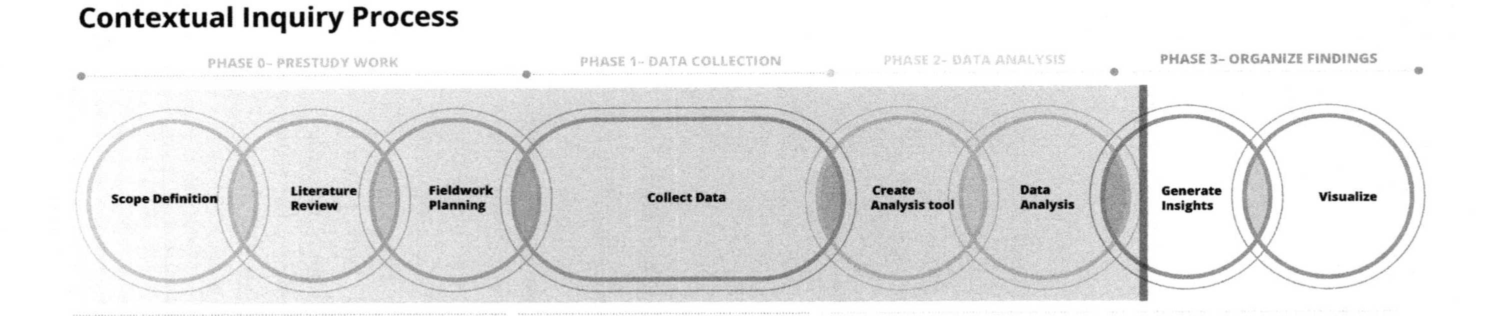

Figure 6.1 Final phase of the CI process is developing design insights.

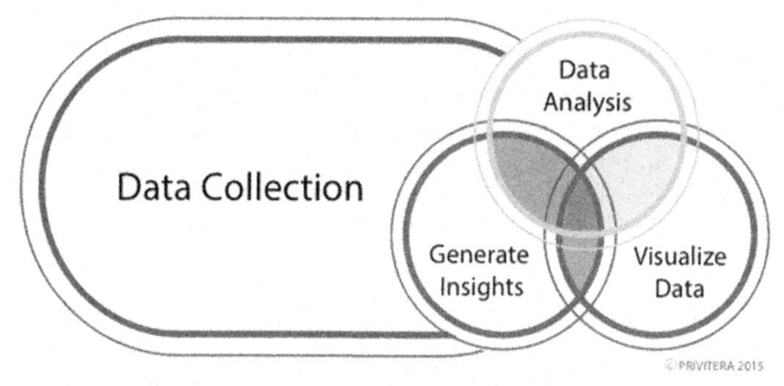

Figure 6.2 Data analysis, generating design information, and data visualization are interrelated.

By understanding why things do not work, you can figure out how to design them so that they do.

This chapter focuses on translating and interpreting the collected data into design considerations, design insights, and ideas. It also discusses user needs as viewed by design control documentation and the practice of writing need statements. Lastly, a quick means of prioritization is provided.

Of note, throughout this chapter there are references to data visualization and analysis. It is difficult to understand what you cannot see. The team must invest in the process of blending data analysis and visualization (Figure 6.2). How the data is compiled makes a difference in the impact it can make. Properly structuring the information facilitates a deeper analysis, and creating dynamic visual output facilitates broader communication and education. When the CI is done properly, interpretation of the data goes beyond the initial insights and becomes a living document; rather than being a one-time activity it becomes a tool for ongoing learning.

6.1.1 Overview

CI requires a state of mind that is open to imagine a new reality. The process of generating design insights begins with internalizing problems, challenges, and behaviors of the user then determining what relevance this has to the product development goal or study goal. It is based on taking the information gathered and processing it to the point that it makes sense in the eyes of the study participant; the product user is the ultimate assessor of the accuracy of the analysis.

"Product developers make incorrect assumptions about user needs, and marketing personnel make incorrect assumptions about the needs of the product designers. Incorrect assumptions can have serious consequences that may not be detected until late in the development process. Therefore, both product developers and those representing the user must take responsibility for critically examining proposed requirements, exploring stated and implied assumptions, and uncovering problems."

(US FDA, 1997)

The above passage from the FDA advocates that determining user needs is the co-responsibility of both the development organization and the user themselves. As such, the development of insights should be done as a team including the primary researcher who orchestrates the work; the product design associates who need the insights to be effective, and the study participants who have a vested interest in the advancement of the product and procedure.

The work is best completed when viewed as an unrestrained accumulation of discovery. While patterns of common discovery help establish priorities for action, often it is the less visible nuggets of data that create insights that lead to breakthroughs. The team should avoid fixating on the obvious few ideas and force themselves to generate a surplus of opportunities for improvement throughout the process.

The interpretation of the collected data assigns meaning about the work structure and possible supporting systems (Holtzblatt and Beyer, 2013). Since design opportunities rely on accurate accounts and interpretations of the research, sharing and reviewing the discovered information is an important step especially since most research teams will never have the opportunity to perform the procedure or use the device studied in a live patient. This is not always easy when discussing the challenges found in the practice of medicine. For example, patients who are critically ill with increased intracranial pressures (swelling in the brain) have no choice for monitoring and will have a catheter inserted several inches into their brains for close watch of this life-threatening situation. In discussion with neuro-critical care physicians they say, "I don't care if it requires the patient to wear a bucket on their head, if it non-invasively monitors pressures then I'd take it!" They do not literally mean they do not care the patient must wear a bucket. Rather, they are passionate about the root need: non-invasive intracranial pressure monitoring. If a design team took a literal interpretation, they would quickly find out the bucket is less than ideal in this use scenario. As upon further examination of patient care there are typically two other common devices being used at the same time as the monitoring unit: a ventilator with an endotracheal tube exiting the patient's mouth and electroencephalography (EEG) leads coming off the head of the patient. They would also discover the patient is being videotaped for the detection of seizure. A bucket is a bad idea. In medical devices, the interpretation turns a fact, an observable event or quote, into a design insight, recommendation, or idea but it requires putting the pieces together and true understanding.

> *"Teamwork is the ability to work together toward a common vision. The ability to direct individual accomplishments toward organizational objectives. It is the fuel that allows common people to attain uncommon results."*
>
> **—Andrew Carnegie**

Generating insights based on CI data collection should include as many members of the product development team as possible and not just the research team. Each member of the cross-functional team has the potential to add value from his or her own perspective and collectively can generate valid user needs based on the evidence presented in the study. Regulatory associates may bring thoughts around compliance to industry standards. Quality associates may bring perspective on the reliability of the process and how risks are mitigated. Marketing associates may focus on the perceived versus actual value propositions. Engineers may provide perspective on how development test methods relate to actual use. Participation creates a level of ownership in the data, and more importantly creates ownership in the need to deliver a solution. Generating insights takes time and cannot be expected to be complete in one session. The process requires the team to immerse themselves the data collected, seek to understand it, digest it fully, and then with take time for proper reflection.

6.1.2 Definitions

Words matter. Often associates confuse design requirements versus specifications, design considerations versus recommendations, and ideas versus concepts. These nuances may create confusion among the team and limit the effectiveness of data analysis. Dependent on the organization, everyone may have a slightly different interpretation for the early phases of the development process however due to design control mandates, the latter phases need to be more consistent to assure the team meets compliance requirements and their work is readily understood by auditors. Listed below are definitions of each word commonly used in generating insights (Figure 6.3). These definitions begin with what we likely already know and then provide an example for use in medical device development.

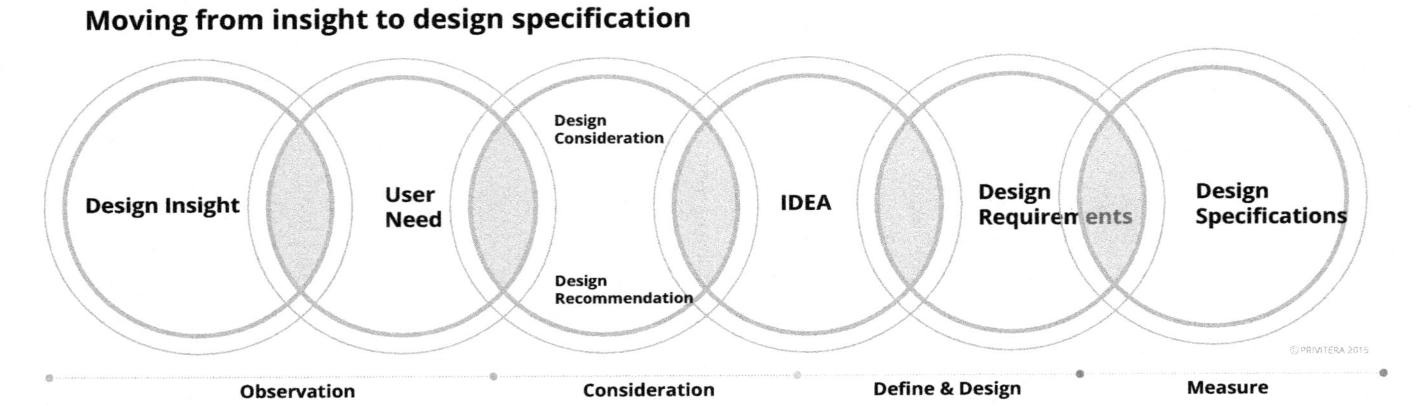

Figure 6.3 Attributes that make up design considerations.

6.1.2.1 Design Insight

Insight (noun): (Oxford Dictionary, 2015a)

1. An understanding of relationships that sheds light on or helps solve a problem.
2. (in psychotherapy) the recognition of sources of emotional difficulty.
3. An understanding of the motivational forces behind one's actions, thoughts, or behavior; self-knowledge.

A design insight is user behavior/s or quotes that are identified as important considerations in designing a new device yet may not be readily actionable. An insight can be a statement made by clinician or an observation made by the team around an issue relating to a particular part of a procedure or device use. Articulating insights are simply a step in the design process for deeper reflection. It encourages you to probe into the "WHY" behind the patterns of the research. Design insights are generated throughout a CI study and require diligence in data analysis to combine what users say with what users do coupled with the realities on the video. It has been suggested that users are not always forthcoming in their behaviors and their verbal suggestions (AAMI, 2014; Privitera et al., 2009; Wilcox, 2010). Therefore insights should be grouped into themes for further understanding prior to writing specific user needs statements.

Example: "As observed, kits are placed on any available surface, including the chest of the patient."

6.1.2.2 User Need

User (noun): (Oxford Dictionary, 2015b)

1. A person who uses or operates something, especially a computer or other machine.

Need (verb): (Oxford Dictionary, 2015c)

1. Require (something) because it is essential or very important.

User needs are straightforward and can be broken down in very specific manners that relate directly to primary functionality (e.g., to drain X or to remove Y). Other needs are more complex and require a deeper understanding of the condition/environment/situation and relate to more subjective factors such as to diagnose, ensure success, prevent, or reduce. These subjective factors often require more information concerning the causalities, mechanisms involved, physiology, etc. before the needs are clearly understood.

Of note, need statements should always be positive statements as it is very difficult to design "not" doing something or "preventing" something from happening. While it is fine to start from this point of view, further analysis into the root cause is required which should enable the formation of a positive statement.

Properly articulating "a need" creates a three-part concise statement of desired outcomes, including:

- a direction (increase/decrease/ensure/avoid),
- a target (comfort, trauma, ease of use), and
- a context (day in the life step/activity).

To be actionable, the "need" must be converted to a "need statement" that is a concise description of a specific goal (what the device needs to do) but does not indicate how to achieve the goal. It may include both quantitative and qualitative targets and use descriptions. For example, a user need for diabetic patients may be the following: "Diabetic patients with neuropathy suffer from foot ulcers. This can be converted to a need statement: diabetic neuropathy patients need a detection means (direction) for ulcer detection (target) during general daily activities (context)." From this statement a design team can conduct further research in each, the direction, target, and context.

> *Example: "Kit packaging should ensure stability while placed on the patient's chest."*

6.1.2.3 Design Recommendation or Consideration

Recommendation (noun): (Oxford Dictionary, 2015d)

1. A suggestion or proposal as to the best course of action, especially one put forward by an authoritative body.

Consideration (noun): (Oxford Dictionary, 2015e)

1. Careful thought, typically over a period of time.
2. A fact or a motive taken into account in deciding or judging something.
3. Thoughtfulness and sensitivity toward others.

Design recommendations or considerations are those high-level attributes that establish a general need but require more specific statements to establish a full set of measurable criteria. For example, from a CI study it may be determined that design considerations for interventional radiology is that the device design should consider that the user will interact with the device in both a light and dark room. Until the core functional concept is flushed out in a particular direction there is nothing else to be done with this fact. Once a direction is settled, this fact can be

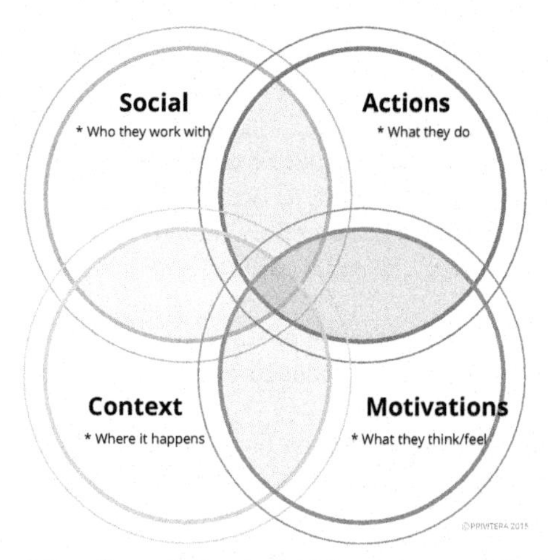

Figure 6.4 The translation from design insight to design specification.

brought out to improve usability and drive dialogue around more specific needs that must be addressed to the use environment. For example, this may be used to determine screen location to minimize glare when the lights are on and include a brightness adjustment to accommodate the changing environment.

There are four main areas of design consideration (Figure 6.4):

- *Social relationships*—this refers to the relationships between stakeholders, how they interact, and overall attitudes toward one another.
- *Specific actions*—this refers to the discreet tasks being performed such as opening a package, handing a device to clinical teammate, etc.
- *Motivations of device use and/or the procedure*—this refers to the data that drives clinical decision making.
- *Overall context of use*—this refers to the environment of use.

The social attributes, actions, and motivations all play a role in the establishment of context (WHO, 2010). An example for each type of design consideration is given in Table 6.1.

This list of attributes is not exhaustive nor is it exclusive, rather the four selected are typical considerations found in a CI study.

Many subtle details are discovered in the process of conducting a CI for medical device development. Having the results serve as a reference throughout the development cycle will maintain traceability and assist in building a robust human factors dossier for agency submission.

Table 6.1 Common Attributes Found in Medical CI Studies

Consideration Type	Example
Social	Design should consider that the procedure would be completed independently; there is no assistance available to users.
Actions	Tying sutures is a fluid process to surgeons; therefore design should consider smooth movements in order to be competitive.
Motivations	Transplant surgeons are broadening their scope of surgical procedures to include gastric banding in order to improve patient outcomes.
	Users have a general aversion to trying unproven technologies or those lacking a strong evidence base.
Context	New devices should network with existing legacy systems and not require special stand-alone equipment to operate.

Examples:

Nurses or techs will prepare the kit then a physician will perform procedure; there are multiple stakeholders (Social consideration). Design should walk the user through set-up and place tools in order of use (Design recommendation).

Kit will be placed haphazardly on patient's chest (Action consideration). Kit design should conform to patients' chest (Design recommendation).

Nurses will be in a hurry during set-up (Motivation consideration). Kit design should self-stabilize once placed (Design recommendation).

Kit will be used in ICU patient room where lighting is low (Context consideration).

6.1.2.4 Ideas or Concepts

Idea (noun): (Oxford Dictionary, 2015f)

1. A thought or suggestion as to a possible course of action.
2. An opinion or belief.

Concept (noun): (Oxford Dictionary, 2015g)

1. An abstract idea.
2. A general notion.

Ideas can include novel concepts that have been communicated either verbally, written in words, or sketched. The ideas can be further defined to potential product embodiments. They require a definition and strategic plan.

Examples:

1. *Curve the backside of the kit to conform to natural chest anatomy.*
2. *Put stilts on the kit and make it hover over the patient like a bedside table.*
3. *Provide straps on the kit that attach to the patient bed.*

4. *Put a conforming foam pad on the back of the kit.*
5. *Make the base of the kit flexible enough that it conforms naturally and then add a side lip around the kit to prevent tools from slipping out.*
6. *Make kit a roll up kit like a craftsman kit.*

6.1.2.5 Design Requirement

Requirement (noun): (Oxford Dictionary, 2015h)

1. A thing that is needed or wanted.
2. A thing that is compulsory; a necessary condition.

Design requirements are the functional attributes that enable the team to convert ideas into design features. From the design requirement statement, a product development team can research all the necessary demographic, anthropomorphic, anatomic, and physiologic data necessary to start developing the actions (design requirements) the device must perform.

Examples: (for this example, the idea "roll up kit like a craftsman kit" was selected)

The kit must contain the following tools: sterile syringe, introducer needle, stopcocks, wire, and exchange catheter.

The kit must be made of a flexible material.

The kit must tolerate ethylene oxide sterilization.

6.1.2.6 Design Specifications

Specification (noun): (Oxford Dictionary, 2015i)

1. An act of describing or identifying something precisely or of stating a precise requirement.
2. A detailed description of the design and materials used to make something.
3. A standard of workmanship, materials, etc., required to be met in a piece of work.
4. A description of an invention accompanying an application for a patent.

The processes of moving design requirements into the final measureable explicit technical requirements yield the design specifications. These specifications must exist within the confines of the use environment. They must break down the design requirement to its most basic level: each specification must stand on its own and must be tied to a design requirement. Thus the development process is traceable. As a result all design specifications should be measurable, have meaning and a source as to why it is there and be associated with a test methodology.

While design specifications are far more downstream in the device development cycle it is important to note that a CI study will generate data in the form of the constraints for the solution. These can include:

- Patient presentation (anatomy and physiology)
- Patient's surroundings, that is, hospital or home use
- Who is the user? Patient themselves or clinician
- Is there a cost target?
- Are there other solutions available?

Examples:
The kit body will be made of XXXX that conforms to these standards:

1. *ASTM C920; Standard Specification for Elastomeric Joint Sealants*
2. *ASTM C1193; Standard Guide for Use of Joint Sealants*
3. *ASTM D882; Test Method for Tensile Properties of Thin Plastic Sheeting*
4. *ASTM D1117; Standard Guide for Evaluating Non-woven Fabrics*
5. *ASTM E84; Test Method for Surface Burning Characteristics of Building Materials*
6. *ASTM E96; Test Method for Water Vapor Transmission of Materials*
7. *ASTM E1677; Specification for Air Retarder Material or System for Framed Building Walls*
8. *ASTM E2178; Test Method for Air Permeance of Building Materials*
9. *ASTM E2357; Standard Test Method for Determining Air Leakage of Air Barrier Assemblies*

In summary, all design development builds off of the design insights gathered in a CI study. Figure 6.5 illustrates the process of transition and highlights the need for design requirement development to revisit the design considerations previously determined. This assures all factors encountered by the user are accounted for in the product design.

In the transition of design definition, the following further highlights the relationships in between:

- *Insights inform user needs.*
- *User needs drive design recommendations, considerations and design requirements.*
- *Design considerations are descriptive attributes that form the details in design that can improve the device usability.*
- *Concepts are required in order to develop design requirements and specifications.*
- *Design requirements collectively form an idea. Each idea/concept has different design requirements.*
- *Design specifications are refinements of measureable requirements.*

6.1.3 Process of Generating Insights

The relevance of the information collected for new product development is the SO WHAT. Answering the SO WHAT is the "aha" moment. These

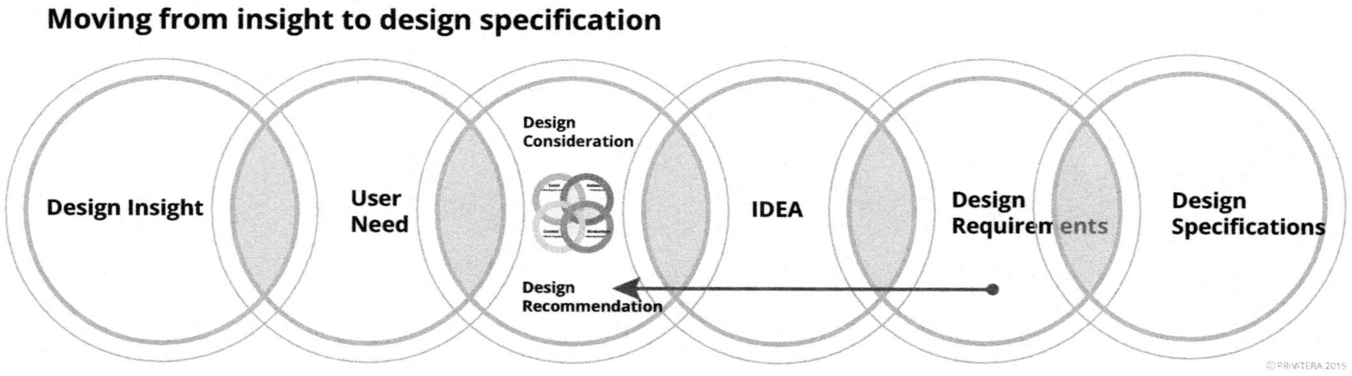

Figure 6.5 CI data analysis to synthesis diagram.

are the discovery of insights. This is a sudden comprehension that solves a problem, reinterprets a situation, explains a joke, or resolves an ambiguous percept (Kounios and Beeman, 2009). An insight occurs when people recognize relationships or make associations between objects and actions that can help solve new problems (Encyclopedia Britannica, 2011). Insights help uncover and reframe the true nature of a problem. Good insights that are based on good research helps to know what the customers' needs are even perhaps before they communicate it. Figure 6.5 below illustrates how information is collected in the field, is analyzed, and then combined and collated in visual manners in order to further determine design information.

As discussed in previous chapters, the field data is coded and grouped. Once this is complete, further analysis can be completed using theme and CI maps. Table 6.2 includes areas of further exploration and refinement for the generation of design information.

Each important piece of evidence can generate a design insight. This involves selecting the most appropriate representative of evidence from the data analysis and determining the specific challenge, the mitigation, and subsequent set of design insights and recommendations (Figure 6.6). For example, in the user quote below regarding the use of a kit to perform a procedure:

> *"In a perfect world you'd set the thing down, open it up like a book. Everything would be laid out in the sequence you need to use it. It would be everything you'd need to make the field appropriate."*

Table 6.2 Prompts for Generating Themes and/or Insights

For Data Combination and Correlation, Try This:

- Review all the data generated: the themes, specific observations, challenges, and mitigations during device use.
- Generate statements on the obvious behaviors or opinions demonstrated. Then look beyond the obvious.
- Reframe the problem.
- Look for connections between behaviors and opinions.
- Look for connections between product design and use behaviors.
- Look for connections between product design and clinical goal.
- Find analogous technologies or situations.
- Define various user types.
- Review all the data generated: the themes, specific observations, challenges, and mitigations during device use.

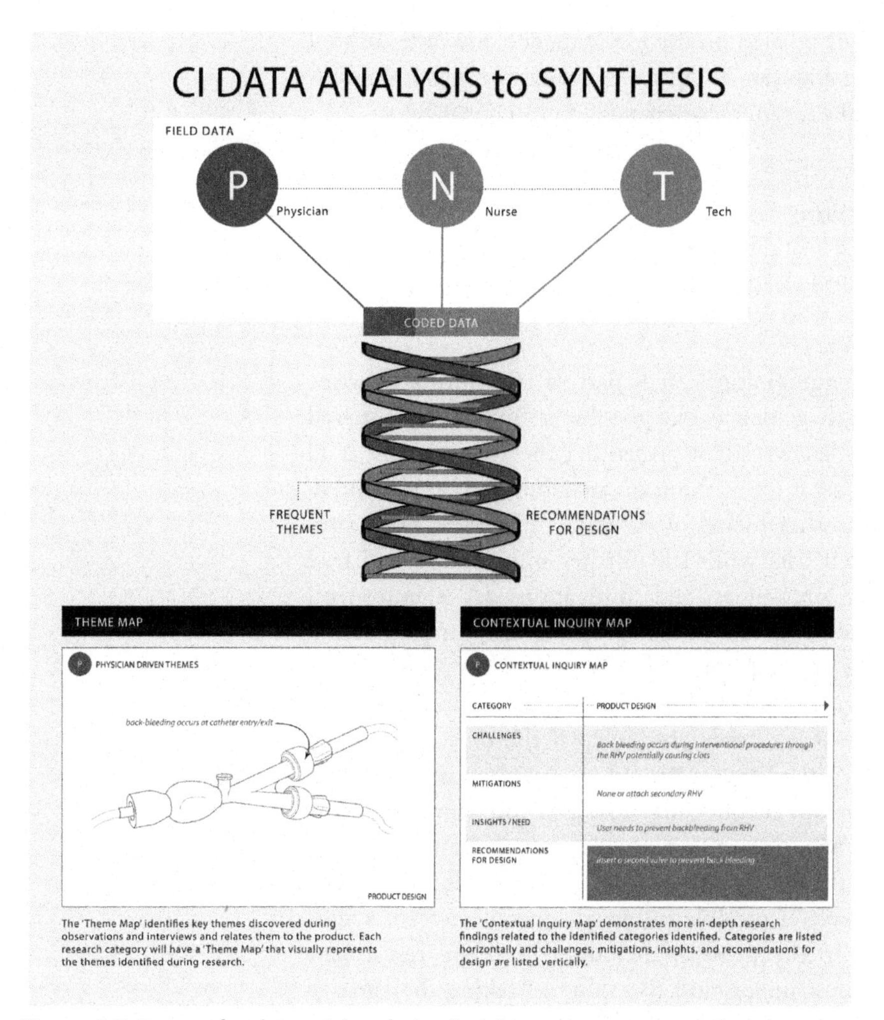
CI DATA ANALYSIS to SYNTHESIS
FIELD DATA
P
Physician
N
Nurse
T
Tech
CODED DATA
FREQUENT THEMES
RECOMMENDATIONS FOR DESIGN
THEME MAP
P
PHYSICIAN DRIVEN THEMES
back-bleeding occurs at catheter entry/exit
PRODUCT DESIGN
The 'Theme Map' identifies key themes discovered during observations and interviews and relates them to the product. Each research category will have a 'Theme Map' that visually represents the themes identified during research.
CONTEXTUAL INQUIRY MAP
P
CONTEXTUAL INQUIRY MAP
CATEGORY
PRODUCT DESIGN
CHALLENGES
Back bleeding occurs during interventional procedures through the RHV potentially causing clots
MITIGATIONS
None or attach secondary RHV
INSIGHTS / NEED
User needs to prevent backbleeding from RHV
RECOMMENDATIONS FOR DESIGN
The 'Contextual Inquiry Map' demonstrates more in-depth research findings related to the identified categories identified. Categories are listed horizontally and challenges, mitigations, insights, and recomendations for design are listed vertically.

Figure 6.6 Process for determining design insights and recommendations based on evidence.

Following the above evidence (quote), the challenge and mitigation is described below.

Tools are not readily available when they are needed	User must search for specific tools taking extra time.
Challenge	Mitigation

Using a full procedure map based on the use sequence, the order of tool use can be determined. In addition, those tools not used can be removed from the kit content list and a design recommendation can be written:

Design Recommendation:
Layout of the tools inside the kit should be in this order: sterile drape, wipes, scalpel, swabs, hemostat, sterile saline syringe, packing material, and scissors.

This one insight/challenge /mitigation is part of an entire CI procedure map. It only points to evidence that the kit is improperly ordered but does not indicate in what ways is it presenting the wrong tools, if all tools are required, and in what order the tools may need to be presented. Coupling this quote with other pieces of evidence, interview transcripts and images from the coded data will yield further information and lead into full descriptions of challenges and mitigations. A theme map (Figure 6.7) can be used to bring like information together in preparation for a full CI map. Theme maps are most helpful when the design team needs to revisit a particular issue.All of the individual theme maps form the larger CI map. The goal is to condense what is the most compelling and truthful, analyze why, write simple statements, and then finesse them.

This process takes time and at times may feel as if the team is splitting hairs. The more diligent the data collection, the more inevitable there will be some level of frustration among the team; however, the persistence to understand as deeply as possible outweighs the short-term aggravations. Each participant in the process, including both the researchers and development team, must understand the value of taking the time to thoroughly understand the data in front of them. The team must avoid "functional fixedness" and seek objectivity. Functional fixedness is the typical bias we all experience when we are "fixed" on the function that is typical, this limits our ability to view data in another sense (German and Barrett, 2005). It is the "its always been done this way" syndrome that can make cognitive road blocks for creative solutions to emerge.

6.2 REVIEW, SHARE, AND ORGANIZE

Throughout the process the team members should be sharing information, and they should seek alternative perspectives that differ from

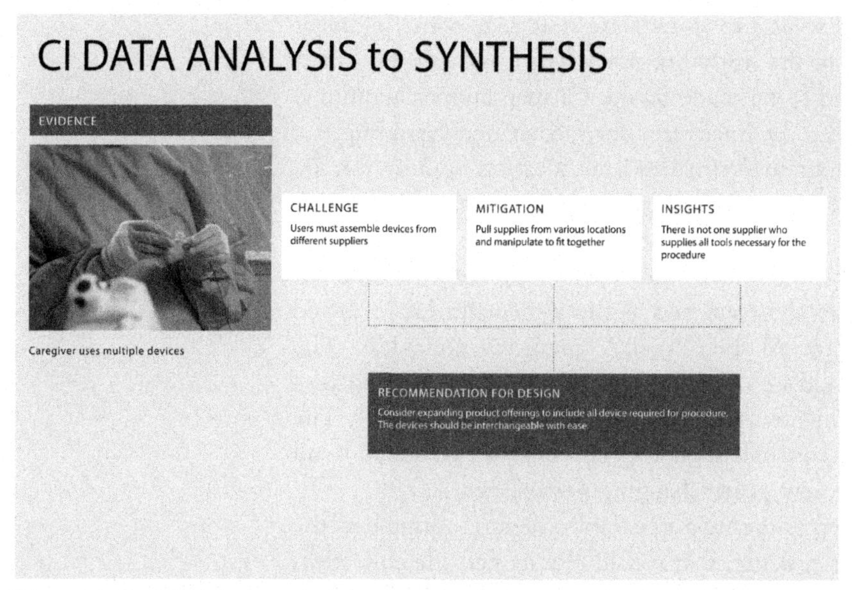

Figure 6.7 Details for generating theme and CI maps.

their own. This is the opportunity for all cross-functional team members, regardless of study participation, to have an opportunity to ask further questions concerning a particular topic or event in interpretation sessions. This review can uncover challenges in communication, opportunities to mine the data further and generate further insights. In this process a team develops an understanding through mutual inquiry into the meanings and facts about their users' work (Beyer and Holtzblatt, 1998). The project benefits from diversity of perspective on the raw data, but ultimately the analysis process should help create a more cohesive perspective on the ultimate design specification that will provide value to the user.

In review sessions, study participants and/or other clinical providers should reflect on the information gathered as it is organized in various visualizations. It is helpful for the clinical providers to review and assure accuracy of nomenclature and interpretation of behaviors. In the review sessions they should be free to elaborate on any point, thus providing deeper meaning, however, they cannot change what was seen or heard.

The goal of these review sessions is to hammer out various possible interpretations and extract meaning that accurately reflects the user and is justified by evidence produced in the research. This practice typically occurs when the majority of the data has been collected and analyzed and serves as an initial creative session to summarize results. In some instances this is done near the end of the study but not when it is finalized. Often there are updates that need to be made to the CI map and/or additional information that may need to be uncovered for further understanding.

Once the research program is finalized an ideation session can be organized around prioritizing needs and generating possible solutions. A review of the disease state, anatomy, and physiology can be used to prime the team for a review of the research findings. Post-it notes of different colors can be passed out to the team and as the information is reviewed they should be asked to record their "aha's," questions, and ideas. The "aha's" and questions should be discussed and the needs prioritized in a simple manner using the mantra: step, stretch, and leap (Figure 6.8). The team may generate partial concepts such as details that can be added to an existing device or entirely new game-changing possibilities.

Figure 6.8 defines a step as the next generation design, a stretch as the one after that, and a leap is a design that would require considerable effort to commercialize (Huthwaite, 2007). By having the entire cross-functional team present, immediate reactions can be provided. This quick judgment will need more thorough analysis but can be useful to direct a focused ideation session. In a focused ideation session, the team starts from their description of ideal then works back to the existing devices. In this approach, working back from the ideal future solution frees the team from the constraints of the present (Huthwaite, 2007), thus providing new opportunities and strategies for development (Figure 6.9).

6.3 BEST PRACTICES

- *Rely on experience*—While it is important to avoid unproductive bias, there is great value to leveraging experience to allow the research to target key areas and to develop more definitive study protocols. The more practice at writing insights and user needs the better. If there is an area that needs more exploration then take it on before the product design is too far along in the development process and changes become problematic.
- *Timing*—Insights are generated throughout the research, they happen at any phase. When they "pop" into a team member's head, they should be noted and shared. Many times this happens during an unrelated event or in discussion with another member of the team. If they are not written down and maintained in a log, they may get lost.

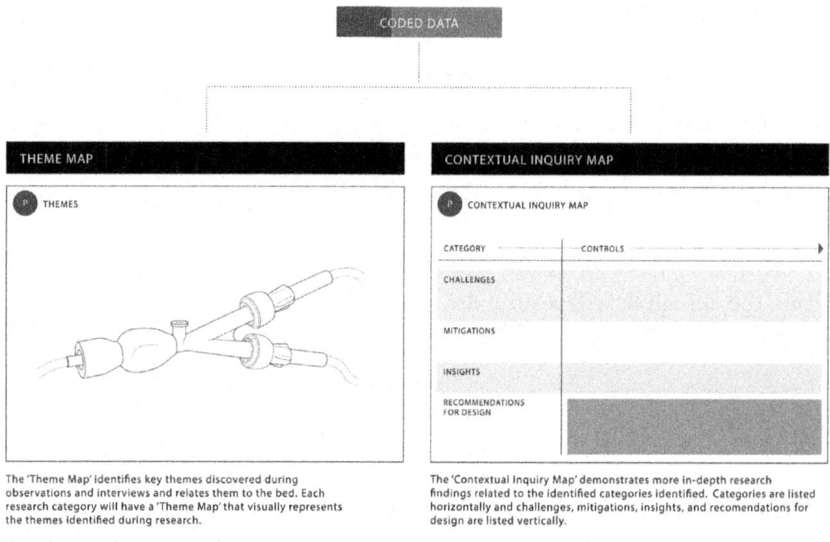
CODED DATA
THEME MAP
THEMES
CONTEXTUAL INQUIRY MAP
CONTEXTUAL INQUIRY MAP
CATEGORY
CONTROLS
CHALLENGES
MITIGATIONS
INSIGHTS
RECOMMENDATIONS FOR DESIGN
The 'Theme Map' identifies key themes discovered during observations and interviews and relates them to the bed. Each research category will have a 'Theme Map' that visually represents the themes identified during research.
The 'Contextual Inquiry Map' demonstrates more in-depth research findings related to the identified categories identified. Categories are listed horizontally and challenges, mitigations, insights, and recomendations for design are listed vertically.

Figure 6.8 Prioritize ideas.

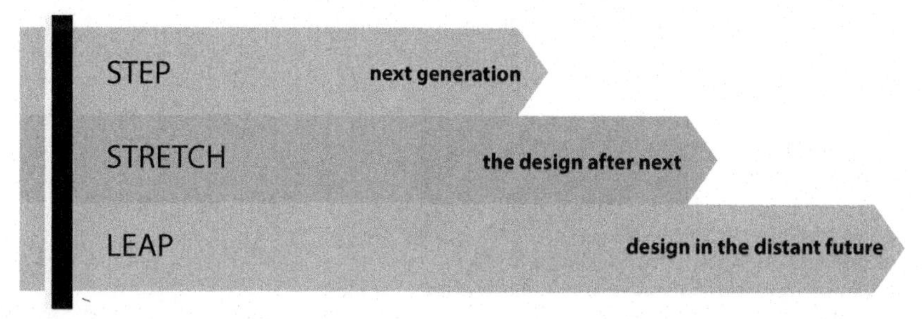
STEP
next generation
STRETCH
the design after next
LEAP
design in the distant future

Figure 6.9 Step-Stretch-Leap, a quick call for identifying the amount of effort or perception of an idea.

- *Individual + Team*—There is value in both individual perspective and collective teamwork. Each person on the team has a unique perspective and will view the data through their own particular lens. Developing design information is a team sport but only when the players come prepared. Taking the time to take a deep dive into the data collection, analysis, visualization, and writing design information will assure that each team member has a shared vision of the ultimate medical device design.
- *Repetition*—Once you review all the data, do it again. Once you generate an insight, read it, challenge its validity and discern phrasing, then trust your gut. The team may challenge it or it may need to be

combined with other insights that are like it. There is not necessarily a right or wrong in writing design information however there is a better or worse.

- *Organization*—Collect all the design insights and recommendations and group them into theme maps. Then select ideation session content to be focused on those themes that are the most difficult problems to solve. Pulling together a large medical device development team for the purposes of ideation is expensive. Focusing the session on particular themes generated from the research will assure the most difficult problems get solved with the broadest input possible.

REFERENCES

AAMI, 2014. AAMI TIR51:2014 Human factors engineering—guidance for contextual inquiry.

Beyer, H., Holtzblatt, K., 1998. Contextual Design: defining customer-centered systems. Morgan Kaufmann.

Encyclopedia Britannica, 2011. Insight: definition of insight in Britannica Online Encyclopedia. Available at: <http://www.school.eb.com.au/all/comptons/article-9275557?query=martin luther& ct=null> (accessed 30.01.15).

German, T.P., Barrett, H.C., 2005. Functional fixedness in a technologically sparse culture. Psychol. Sci. 16, 1–5.

Holtzblatt, K., Beyer, H.R., 2013. Contextual Design. In: Soegaard, M., Dam, R.F. (Eds), The Encyclopedia of Human-Computer Interaction, second ed. Aarhus: The Interaction Design Foundation.

Huthwaite, B., 2007. The Lean Design Solution: A Practical Guide to Streamlining Product Design and Development. Institute for Lean Innovation, Michigan, USA.

Kounios, J., Beeman, M., 2009. The aha! moment: The cognitive neuroscience of insight. Curr. Dir. Psychol. Sci. 18, 210–216.

Oxford Dictionary, 2015a. Insight: definition of insight in Oxford dictionary (American English) (US). Available at: <http://www.oxforddictionaries.com/us/definition/american_ english/insight> (accessed 30.01.15).

Oxford Dictionary, 2015b. User: definition of user in Oxford dictionary (American English) (US). Available at: <http://www.oxforddictionaries.com/us/definition/american_english/user?searchDictCode = all> (accessed 30.01.15).

Oxford Dictionary, 2015c. Need: definition of need in Oxford dictionary (American English) (US). Available at: <http://www.oxforddictionaries.com/us/definition/american_english/need?searchDictCode = all> (accessed 30.01.15).

Oxford Dictionary, 2015d. Recommendation: definition of recommendation in Oxford dictionary (American English) (US). Available at: <http://www.oxforddictionaries.com/us/definition/american_english/recommendation> (accessed 31.01.15).

Oxford Dictionary, 2015e. Consideration: definition of consideration in Oxford dictionary (American English) (US). Available at: <http://www.oxforddictionaries.com/us/definition/american_english/consideration> (accessed 31.01.15).

Oxford Dictionary, 2015f. Idea: definition of idea in Oxford dictionary (American English) (US). Available at: <http://www.oxforddictionaries.com/us/definition/american_english/idea?q = Idea> (accessed 30.01.15).

Oxford Dictionary, 2015g. Concept: definition of concept in Oxford dictionary (American English) (US). Available at: <http://www.oxforddictionaries.com/us/definition/american_english/concept?searchDictCode = all> (accessed 30.01.15).

Oxford Dictionary, 2015h. Requirement: definition of requirement in Oxford dictionary (American English) (US). Available at: <http://www.oxforddictionaries.com/us/definition/american_english/requirement> (accessed 31.01.15).

Oxford Dictionary, 2015i. Specification: definition of specification in Oxford dictionary (American English) (US). Available at: <http://www.oxforddictionaries.com/us/definition/american_english/specification?searchDictCode = all> (accessed 31.01.15).

Privitera, M.B., Design, M., Murray, D.L., 2009. Applied ergonomics: determining user needs in medical device design. Conference proceedings: Annual International Conference of the IEEE Engineering in Medicine and Biology Society. IEEE Engineering in Medicine and Biology Society Conference, 2009, pp. 5606–5608.

US FDA, 1997. Design Control Guidance for Medical Device Manufacturers, USA.

WHO, 2010. Context Dependency of Medical Devices Background Paper 5. Available at: <http://whqlibdoc.who.int/hq/2010/WHO_HSS_EHT_DIM_10.5_eng.pdf>.

Wilcox, S., 2010. Ethnographic research for medical-device design. MDDI Medical Device and Diagnostic Industry News Products and Suppliers. Available at: <http://www.mddionline.com/article/ethnographic-research-and-problem-validity>.

CHAPTER 7

Data Visualization and Communication

Mary Beth Privitera
University of Cincinnati and Know Why Design, LLC, Cincinnati, Ohio, USA

Contents

7.1 OVERVIEW

Every medical device development program needs to begin with a set of basic information that allows the team to galvanize on a common objective. Even rough contextual inquiries done very early in the design process can be used for inspiration and provide vision and reference in long and arduous product development processes (Koshinen et al., 2011). Visualizing and communicating the findings acts is the source of this clue and can serve as a constant reminder of the user and the use environment.

Visual mapping techniques of clinical procedures can be a powerful tool to determine unmet user needs as well as other areas of improvement and opportunity in the context of use (Privitera et al., 2012). It can assist a product development team in the development of a comprehensive understanding. Collecting the data is a valuable step, but it is data for data

M.B. Privitera: Contextual Inquiry for Medical Device Design.
DOI: http://dx.doi.org/10.1016/B978-0-12-801852-1.00007-1

sake if it does not facilitate learning. The impact of the data on the team is directly related to the manner in which it is communicated; the depth of overall understanding and the ability to explore the information further increases with the clarity with which each insight is captured. Visual representations, such as mapping, assure communication of the research findings across various disciplines and can be used to confirm design assumptions or interpretations with users directly. These visual maps provide a means of readily communicating outside of the immediate design team and for assuring data are not lost over time. Ideally, these maps facilitate ongoing references that give the team a sense they are immersed in the use environment, a means for checking their work routinely, and for capturing new learning as the program progresses.

This chapter focuses on the communication of findings using proven mapping techniques (Figure 7.1). It discusses the types of data visualizations used, how they are used, implemented, and provides example of illustration with template maps. Lastly, a discussion on use purposes for the maps from multiple perspectives is included.

7.2 TYPES OF DATA VISUALIZATION

Due to the Internet there is an amazing amount of proliferation, collection, and organization of data; in our society connections and relationships are visualized and mapped at an unprecedented rate. Many of the visualizations are practical and aid in understanding relationships within a complex site or organization. Others deal with complex relationships among data and their creation is aided and would be nearly impossible without sophisticated computer applications (Katz, 2012). The fact is we live in a visual-rich world; by producing visualizations of clinical procedures the data becomes more dynamic and impactful.

Graphical displays should:

- *Show the data*
- *Induce the viewer to think about the substance rather than about methodology, graphic design, the technology of graphic production or something else*
- *Avoid distorting what the data have to say*
- *Present many numbers in a small space*
- *Make large data sets coherent*
- *Encourage the eye to compare different pieces of data*
- *Reveal the data at several levels of detail, from a broad overview to the fine structure*

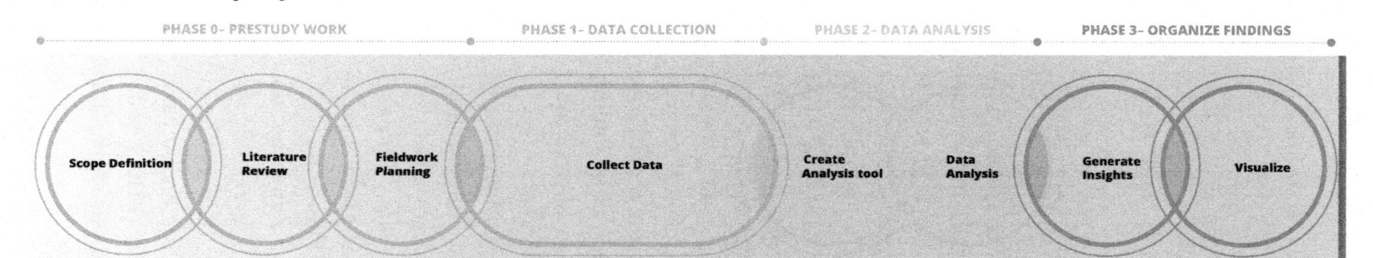

Figure 7.1 Completing final visualizations of information are the final steps in a contextual inquiry (CI) study and enables the transfer of information to device design.

- *Serve a reasonably clear purpose: description, exploration, tabulation or decoration*
- *Be closely integrated with the statistical and verbal descriptions of a data set.*

Graphics reveal data. Indeed graphics can be more precise and revealing than conventional statistical computations (Tufte, 1983).

Likewise, data visualizations for CI studies should include the use of images, illustrations, video, procedure maps, storyboards, and spider diagrams.

7.2.1 Images

Still images provide the most functional means for capturing data. The ability to see both the overall environment and precise details and to capture this information relative to a specific point in the research process is invaluable. These images give the ability to understand interactions and limitations allowing the team to dissect the "WHY" behind the "WHAT" they observe. Often CI studies begin with a less than perfect but deep-rooted understanding of the customer; there are few means more effective at disproving false assumptions than a photograph. Still frames enable you to study a point in time in great detail thus provides the opportunity to pause and study. In a CI study where device interaction with hand movements or device interactions with anatomy are goals, still photography can be the most helpful in analysis as points of interest are readily marked with arrows. Lastly, using the resolution of traditional cameras provides higher resolution than screen grabs from video.

Any imaging in a clinical environment will require consideration of privacy requirements, but this should not preclude photography. Photographic images are best used to illustrate a particular event that occurred during the fieldwork for reference later on. As such, all images gathered in a study will require processing to ensure that study participants are unidentifiable this includes HIPAA-related data (Figure 7.2).

Images should be selected based on which image best represents the theme or user need as determined in data analysis. In Figure 7.2, the challenge of space for sterile bedside procedures in the ICU is clearly demonstrated. The provider must use the bedside table for kit preparation, which leads to drapes obscuring the kit. The team may capture many images that are suitable for describing the same topic; when this occurs all the images should be evaluated and noted before down selecting. There are often subtle details that will suggest one image is stronger than another to communicate a particular attribute.

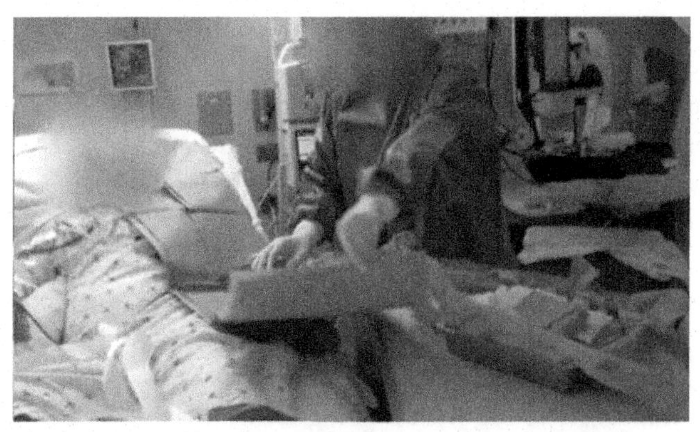

Figure 7.2 Patient and provider faces are blurred to protect identities.

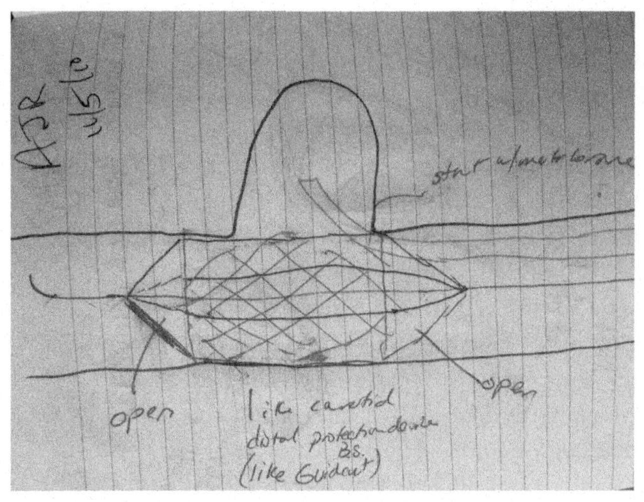

Figure 7.3 Sketch by Dr. Andy Ringer explaining the use of liquid embolic agent for the treatment of intracranial aneurysm.

7.2.2 Illustrations

Illustrations are best used to communicate complexity. Intricacy in medicine is commonplace and medical training routinely uses illustrations to create simple representations of quite complex anatomy, physiology, and clinical application. Illustrations are how doctors learned their science and how they generally communicate with each other and with patients (Figure 7.3). Figure 7.3 is a sketch by Dr. Andy Ringer, a neurosurgeon who is cross-trained in endovascular approaches. In this image the research team handed him their notebook, provided several color pens with the rule of "one color = one part (anatomy or device)." Surgeons in particular commonly draw the anatomy they work on. Illustrations like the one below, can simplify device–tissue interactions especially in complex cases.

The development of graphical representations allows the team to simplify the observation, to distill it down to the critical elements, and to communicate those elements better than a still image. Illustrating device interaction with anatomy can decipher key device and tissue interactions that may or may not be adequately functioning in a procedure. Further, many procedures are minimally invasive and thus it is impossible to see what is going on with the patient's anatomy *vis-a-vis* devices in one view (Figure 7.3).

Anatomical illustrations show the physical appearance of things but they do not show how the body works, how movement affects it. There is a great deal of information to be gained from anatomical drawings and helps plan a procedure, however sometimes the lack of fidelity can actually mask important elements of the subject (Katz, 2012). In medical device development, having an understanding of the anatomy is not enough information. The anatomical illustrations must also have device interactions and are often presented in a series. In Figure 7.4, the simplified illustration of the kidney, ureter with stone, bladder, and the urethra is much easier to discern than the fluoroscopy image provided in the upper left box on the illustration. The study participants, namely the surgeons who often use illustrations to explain their procedure, first initiated this anatomical illustration. From this illustration, it is possible to layer on the devices as they interact with the anatomy in order to understand the order, the significance of surface texture for both adequate access to the stone as well as tissue interactions. Illustrations such as this were used in data analysis and on the final procedure map in a CI study focused on urologic care.

Likewise, illustrations are helpful to communicate the internal mechanisms of devices as they are used (Figure 7.5).

Figure 7.5 is of an RHV used in interventional radiology procedures. The purpose of an RHV is to provide a seal at the entry point of the artery. An introducer catheter is placed initially and the RHV is attached to it. The RHV provides the main working channel for the procedure. It is a staple in these types of procedures. One of the challenges communicated by CI study participants was back-bleeding from the proximal hub connection. This back-bleeding is a cause for concern in that it could be

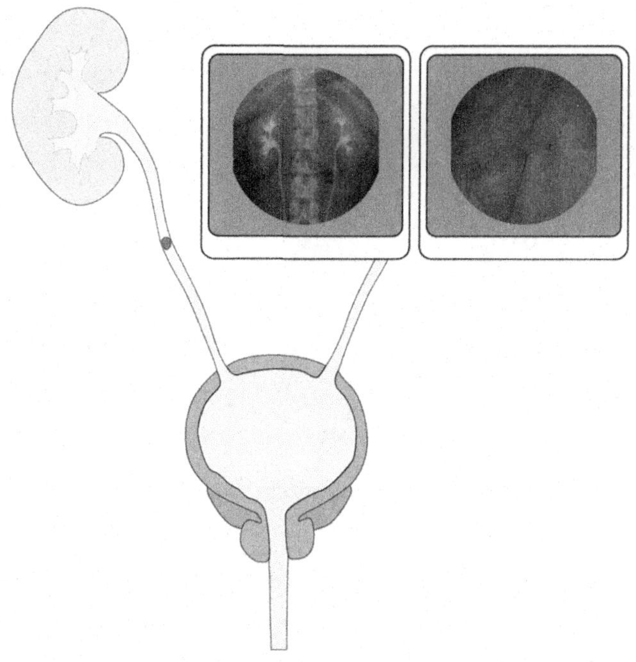

Figure 7.4 Illustration of a kidney stone caught in the ureters coupled with images from fluoroscopy (on the left) and the ureteroscope (on the right).

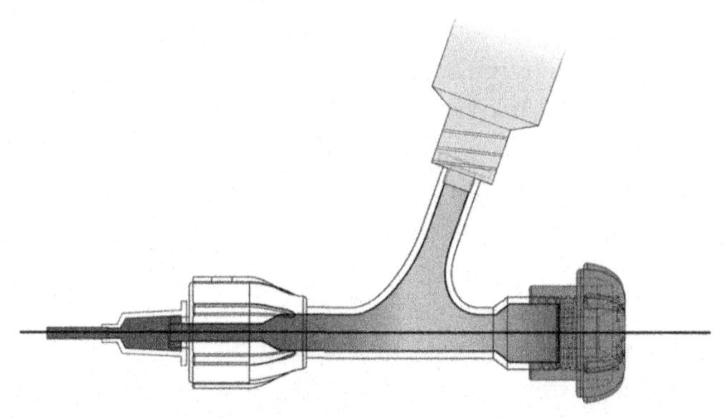

Figure 7.5 Rotating hemostatic valve (RHV) used in interventional procedures illustrating back-bleeding in the chamber.

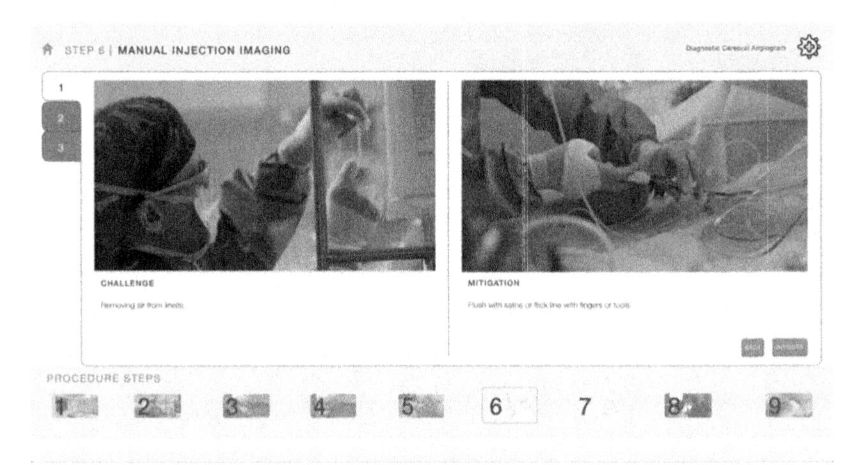

Figure 7.6 User interface of an interactive CI database for cerebral diagnostic angiogram procedures.

the source for an embolization on the catheter or wire that is inserted through the main working channel.

7.2.3 Videos

Video information provides rich and comprehensive data within CI studies. Continuous video allows for far more data collection than other means including recording when researchers cannot be present; this also provides for correlation to physiologic monitors. Video footage is most helpful when conduction motion studies, where nuances of movement, play a role in the design process.

During data analysis videos are typically tagged and coded for significant events. These tags can be selected much like the most representative of a situation being selected. Video clips can be used in PowerPoint presentations and be integral to an interactive database of CI information (Figure 7.6). They provide for broad education of the team, help in the development of simulation models, and are extremely valuable when trying to optimize study techniques.

Figure 7.6 is a screen shot image of an interactive database regarding cerebral diagnostic angiograms. The database is navigated by clicking on each procedure step along the bottom. Each step is explained in the first tab (shown) and then challenges and mitigations are demonstrated in tab 2 followed up with insights in tab 3. When the user clicks on an image a video is played to further illustrate the procedure and/or challenge. This database serves as a training tool for residents and acts as an introduction to interventional cases for engineering.

The major benefit of the interactive database is its easy reference to the video. Videos are clipped to include only the most important elements with text explanations below. This enables a product development team or resident to skip around and focus on only those steps or elements of immediate interest. The downside of the database is that it is impossible to view the entire procedure at once. Thus it is difficult to determine the connections between steps unless explicitly described. Ideally, a product development team has both a large printed CI map and an interactive database.

7.2.4 Procedure Maps

Procedure maps are the final and full visualization of a CI study. They typically include all of the details regarding the number of observations, the location, and the study sponsor. These maps are intended to drive discussion by bringing the most pertinent information together in a manner that facilitates the "aha" moment. The format of the map should be specific to the specific goal, but some elements are common. Communication of sequential relationships is typically recorded horizontally creating discrete steps. Each of those steps can then be analyzed in the vertical plane providing the ability to explore each in depth. A typical exploration begins with a visual image, adds explanatory data, identifies challenges, and concludes with opportunities.

Figure 7.7 is a full CI map based on a cerebral diagnostic angiogram study.

The map highlights each step of the procedure and provides details regarding the task description, the tools used, the challenges, potential mitigations, variations between physicians, and insights for design. Along the horizontal plane the map provides the main procedural steps with standardized descriptions that will act as a means for dissecting the overall workflow for a variety of reporting purposes. The work breakdown process needs to go beyond the obvious; it must capture transitions in both physical actions and cognitive awareness/clinical decision-making. Along the vertical plane the procedure steps are analyzed from general to specific communicating a clear message that stands on its own distinct from the other steps. When developing the map, the team should be cognizant of the many ways it will be used, and attempt to communicate accordingly using consistent terms and self-explanatory images. Implemented, use

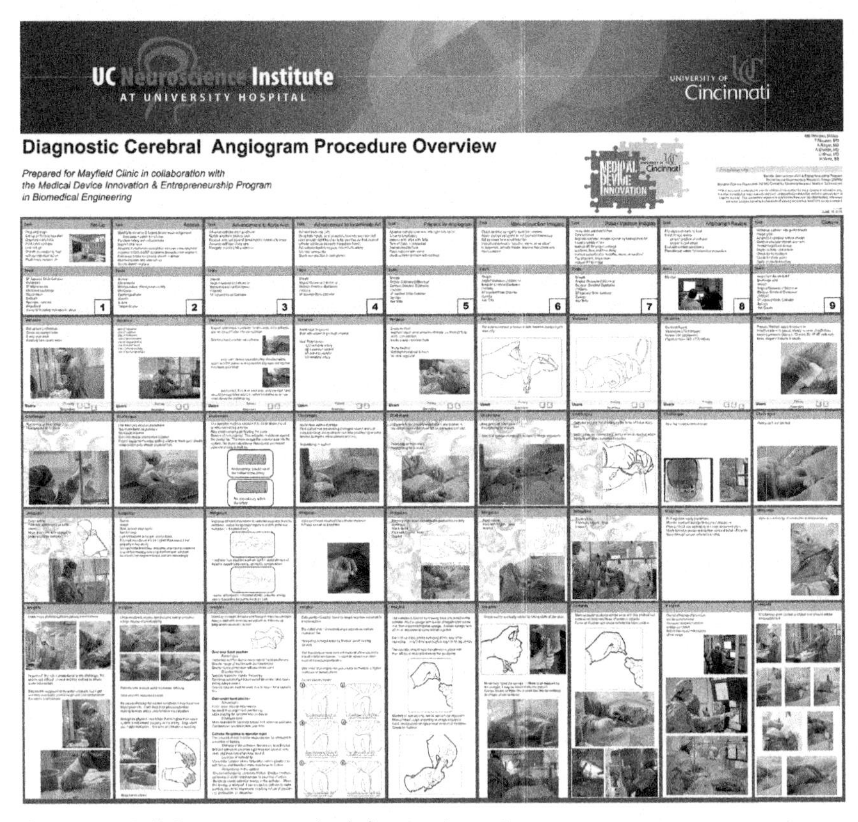

Figure 7.7 Full CI map on cerebral diagnostic angiograms.

challenges and mitigations, variances between physicians, and insights for design. The map is organized to begin with general information at the top and moves to more specific details down the column.

This particular map was generated in collaboration with a large number of users with diverse perspectives; a comprehensive review process, including many of those same users, helped ensure the results were communicated in a manner that facilitated a common interpretation. Embedded in the map are images of bubbles at various places: steps 1, 5, 6, and 7. The removal of bubbles as an activity was of high importance to the clinical team and noted by the research team. The significance of their removal impacted the success of the case by preventing complications such as an air embolism. The research team noted that up to 30 minutes of time is dedicated to this important activity. The bubble image is placed in the background of the steps to highlight the connection that searching

Table 7.1 Generic Layout of Map Content

Study Sponsor **Methods, Authors**

Step 1	Step 2	Step 3	Step 4	Step 5
Task description		→	→	
Tools required 1		→	→	
User involvement 1		→	→	
Challenges 1		→	→	
Mitigations 1		→	→	
		→	→	
Insights 1				

and removing bubbles from lines happens often. This use of an image could easily be an icon and represents significant findings in the data. Additionally, illustrations are used to communicate detailed interaction of the device as it traverses up and around the aorta (see step 4 bottom row).

For detailed information regarding the content of each cell, see Table 7.1.

The procedure mapping process starts with a descriptive task analysis, breaking down each step in the procedure into small details that will act as the horizontal headers in the storyboard. Traditional techniques for work breakdown structure can be used to define the high-level individual steps; the process for creating the points of demarcation is often a means for helping the team members look beyond typically understood steps and to go deeper to think about what nuances that the user is experiencing. Some factors to use when defining the steps include points of clinical decision-making, change of control between users, transition from one tool to another, and points of heightened interest in the function of a specific device. Next, data collection topics specific to the study objective should be agreed upon by the team; these will act as the means for the

Figure 7.8 Story board development.

team to collect and analyze the data and act as the vertical labels for the storyboard. Typical vertical labels may include emotion, opinion, purchase decision, clinical decision, or variances. Wherever possible, adding visuals to the information provides greater context and credibility to the display for further and facilitates greater input and deeper critique in the process.

Figure 7.8 is a storyboard of patient movement within the emergency department from initial triage, second triage, waiting room, to the treatment, and/or discharge or next disposition. Simple line drawings based on images were used in order to communicate the interactions between the providers and the patients. Initial task descriptions were provided under each image. The post-it notes represent relative findings such as interesting points, challenges, and mitigations discovered from the field. This image represents a storyboard in progress. The display is intentional for review with providers in order to gain additional insight.

As mentioned earlier the proliferation of data displays on the Internet is staggering. Below is an image of a spider diagram demonstrating that is used to illustrate relative procedure time by steps (Figure 7.9).

Using spider diagrams to display specific quantitative data points help decipher the relationship. In Figure 7.9 it is clear step 5 takes the majority of time whereas step 2 is the fastest. Procedural time is always a measure in medical device use, especially in the development of surgical tools.

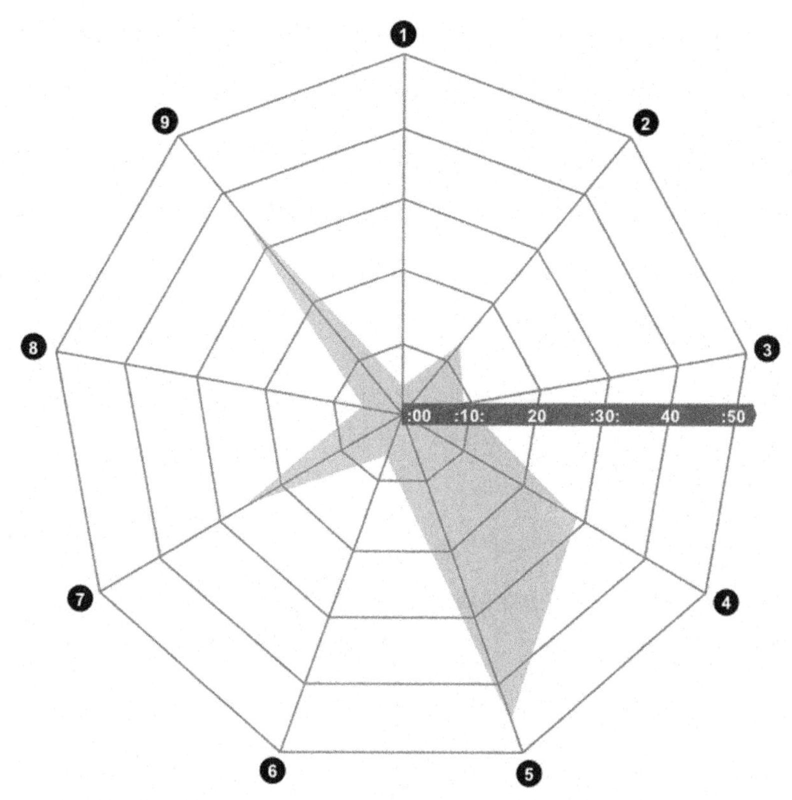

Figure 7.9 Spider diagram demonstrating relative time measures for procedure steps.

The spider diagram alone is not sufficient information as to why a particular step takes longer than another, however a spider diagram coupled with a full map can provide rich data that includes causality.

7.2.5 Procedure Map Templates

There are many variations on building a full procedure map. The selection of design should be based on the goals and objectives of the overall study. Illustrated below are further examples of map templates that highlight variations in study emphasis as well as layout.

Figure 7.10 procedure map template can be used for studies that involve two sites or areas of care. In this format the steps are on the horizontal access and each column contains information regarding specific observations, the challenges, and mitigations then finally the insights. Study details are found on the left with descriptions of each row in the

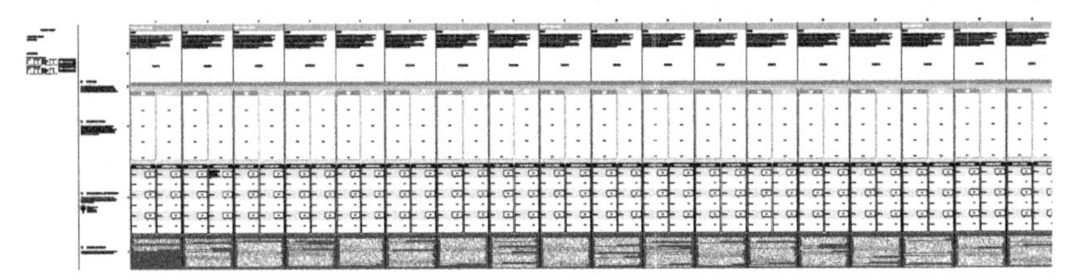

Figure 7.10 Procedure map template for studies with two locations of care.

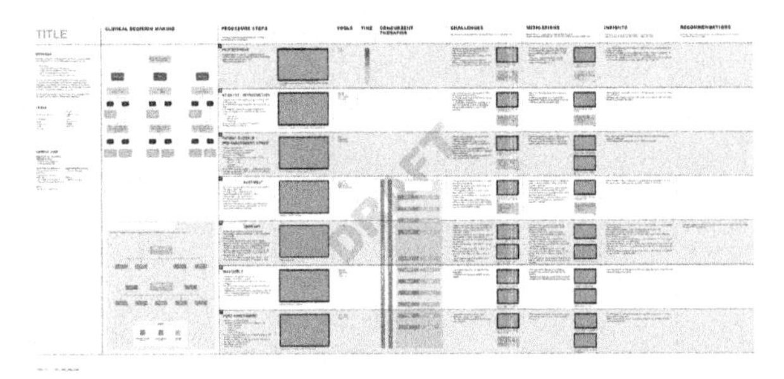

Figure 7.11 Procedure map template with clinical decision-making process diagram and a vertical procedure step format.

next column. This map is more complex and includes the ability to include photographic evidence to support specific challenges and mitigations in the field from two separate locations in comparison.

Figure 7.11 presents an alternative layout with the steps of the procedure in a vertical format. This template also includes a clinical decision diagram prior to the procedure steps. This template is appropriate for procedures that are not overly complex. Having the procedure steps stacked on top of one another poses limitations due to plotter size.

Figure 7.12 combines procedure steps, observations, clinical decision-making, clinical (user) preferences, and insights in one map. In this map each element is plotted with the procedure steps. The inclusion of preference needs context. In this example, it is the positive or negative opinion regarding the procedure step and not necessarily a particular device. Gathering opinion regarding a particular step can assist in the identification of opportunity from the viewpoint of the user.

Regardless of format, it is suggested that the columns and the rows should be clearly identified in an alpha-numeric system for the purposes of review. In this way during review a particular cell is easily referenced. For example, a member of the team can reference cell A-5 and know where on the map the particular data issue is located.

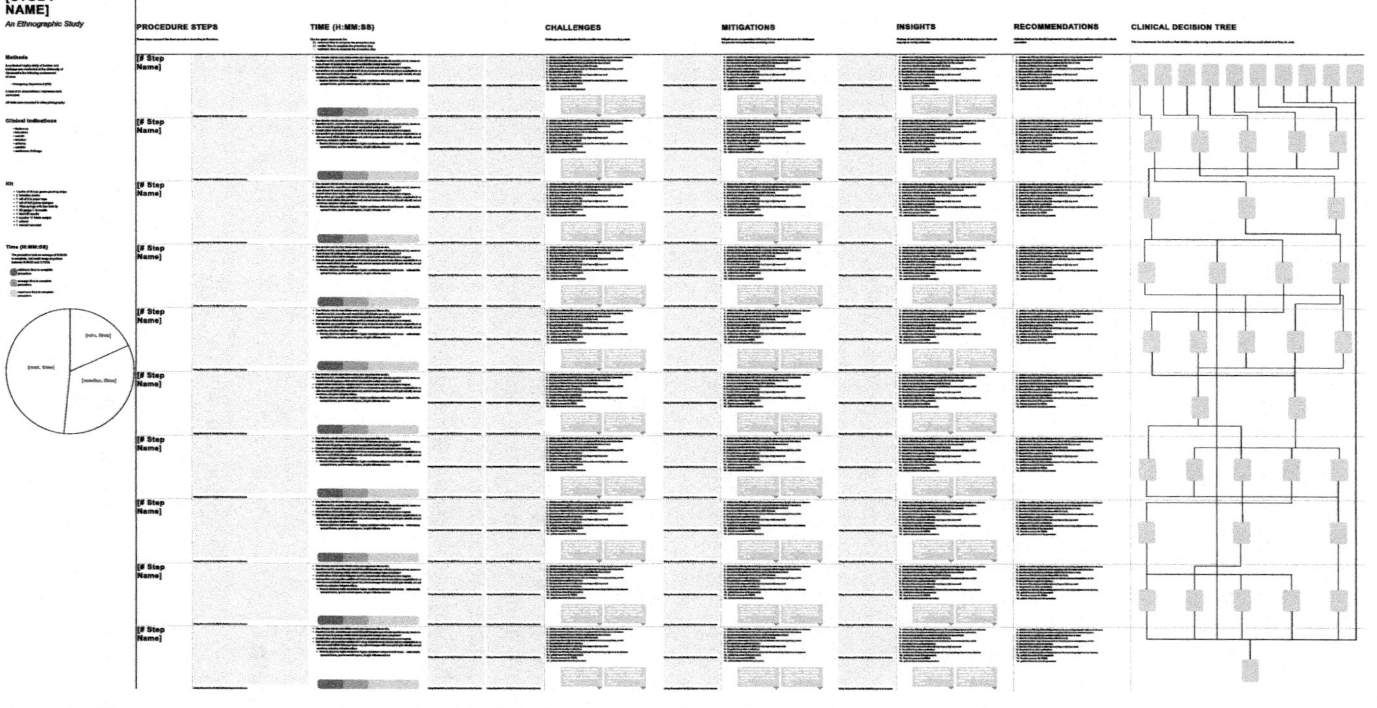

Figure 7.12 Procedure map template with clinical decision and preference for each step.

7.3 COMMUNICATING FINDINGS

7.3.1 Clinical Providers

Wake-up calls are occasionally needed (Koshinen et al., 2011). Thinking that what exists today cannot be improved; many providers readily accept the devices available today as acceptable and meeting their needs. In many instances, they have no other choice, their patients are here now, and they are doing their best with what is been provided. Including providers in map development both confirms interpretation but can also open the participants mind to further seek improvement areas with the research team. In turn, this exchange can further initiate valuable collaboration between the users and the medical device developers.

In addition, maps can be used to gather broad opinion if used at a scientific meeting for reference and development discussion. In this instance, the design-related information may need to be removed to protect intellectual discovery. Once completed, the conversation with (additional) providers can be focused on particular aspects of the procedure using the map as a prompt. From here a product development team can gain an appreciation for subtle nuances and the breadth to which they are practiced.

Maps can also be used for clinical training. Maps contain rich information that can assist a novice to learn the procedure. Many surgical manuals and clinical protocols walk through each step of the procedure steps at a top level then communicate key aspects and indications. These may include a limited number of illustrations however; they often do not include the ability to see an entire procedure at once.

Dependent on how and where the novice is being trained the map can take on different forms such as Figure 7.13, an individual step on one page or as an interactive database (Figure 7.6). The page format is good for reference in printed notebooks and the database as a referential guide.

7.3.2 Product Development Team

The overall benefit of the map for product development teams is the ability to view the entire procedure in its entirety and then take a deep dive on a specific step or challenge. All challenges and mitigations are presented with specific evidence and are traceable in the data analysis. This enables further exploration of a specific topic from the database. In addition, product development team members can begin to identify key opportunities, which meet their strategic objectives, and those challenges

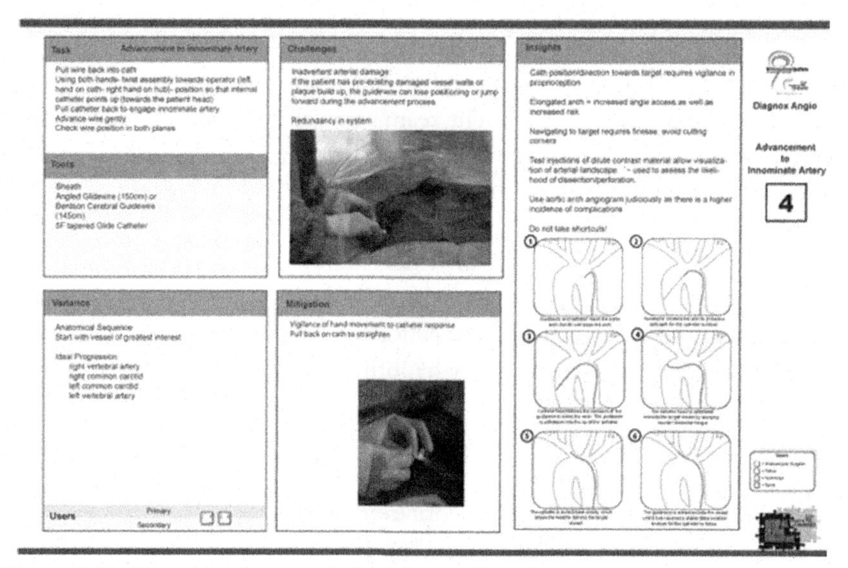

Figure 7.13 Clinical study page derived from a CI map.

that may be important to their users but not necessarily within their current target development criteria.

The visualization of information puts the data in front of the design team so that they can intentionally construct a new future in the form of disciplined imagination (Koshinen et al., 2011). In this manner the map becomes the key means of constructing knowledge. The need statements provided can be grouped and collated in order to inform a new design and innovation. Lastly, further context of use in observed behaviors can serve as a reference to surrounding anatomy and tools used in conjunction of a given step in order to provide.

Maps can also be used to generate conversation with users and engage them in the product development process. The maps can enable conversations on the observations and the possible significant changes through innovation. The map can inform a discussion of potential impact and the overall clinical outcome. In some instances, users may attempt to refute findings. To move past these hiccups, users should be referred to evidential quotes or images.

Lastly, maps are most often displayed near the product development team area and serve as continual reference of context, thereby enabling a reminder of the user experience throughout the development process.

7.4 BEST PRACTICES

- *Continuous review*—Procedure maps should be reviewed as they are being developed. The entire research and development team, as well as key study participants, should actively contribute to the drafting and refinement of every element in the map. The map should be seen as a living document that develops and then evolves.
- *Create references*—Label the data on visualizations or provide citations that allow the research team to identify where to find the raw data. This enables the team to go back and seek further clarification if necessary. Label what was observed directly and what is a hypothesis by the research team.
- *Confirm all the use intentions by the team*—Depending on where and how the map is to be used, more or less information may be desired. For example, if presenting the map at a scientific conference the design insights may need to be removed for protection of product development direction. If the use is intended for clinical education, then multiple means of presentation may be necessary.
- *Keep a list of possible illustrations during data collection and analysis*—Illustrations are used to simplify task descriptions. Having a graphic designer involved in the layout of information throughout the study is helpful. At minimum, they can begin to illustrate steps in the procedure and the anatomy. As the data analysis unfolds more details can be added.
- *Plot many iterations prior to finalizing*—Procedure maps can get quite large and difficult to edit or see trends if left in digital format. Do not hesitate to plot a black–white version and mark it up. This is sometimes a faster means of editing as well as data analysis since the entire map is displayed and in a readable form.

REFERENCES

Katz, J., 2012. Designing Information: human factors and common sense in information design. John Wiley & Sons.

Koshinen, I., Zimmerman, J., Binder, T., Redstrom, J., Wensveen, S., 2011. Design Research through Practice from the Lab, Field, and Showroom. Elsevier, Waltham, MA.

Privitera, M.B., Abruzzo, T., Ringer, A., 2012. Visual mapping of clinical procedures using ethnographic techniques in medical device design. In: Naidoo, L. (Ed.), An Ethnography of Global Landscapes and Corridors. InTech, Croatia, pp. 223–232.

Tufte, E., 1983. The Visual Display of Quantitative Information. Graphics Press, Chesire, CT.

Case Study Introduction

The next few chapters present case studies and stories from professionals who have been involved in conducting contextual inquiry (CI) studies since the early 2000s. They focus on the why and how to conduct a CI study for medical device development. Each case presents the same basic methodology that has been applied in various locations of care and for various purposes. The intent is to provide real examples from practice and present interesting situations and challenges in conducting a CI study.

Each of these case studies presents different approaches to conducting a CI study. They demonstrate the flexibility of the methods and the adaptability of the research team and establish that there is no right or wrong way to conduct a study. Rather better or worse and the study must be carefully crafted for each product development phase and clinical situation.

Many thanks to those who permitted the use of the following case studies: the sharing of processes is vital toward advancing medical device development practice. Their contribution is greatly appreciated.

A CI STUDY IN CEREBRAL ANGIOGRAPHY FOR PRODUCT DEVELOPMENT STRATEGY AND CLINICAL TRAINING *MARY BETH PRIVITERA & ANDREW RINGER, MD*

Chapter 8 is a CI case study on diagnostic cerebral angiogram conducted at the University of Cincinnati (UC), USA. Diagnostic cerebral angiograms are an interventional radiology diagnostic procedure performed to assess potential cerebrovascular health or blood flow in the brain. During the procedure, a catheter (thin, long tube) is inserted through the femoral artery at the groin and then navigated through the patient's vascular system. The goal of this program was to work collaboratively with the Department of Neurosurgery in order to identify key behaviors and actions that are attributed to challenges in the procedure, identify new product development opportunities, and lastly develop training materials for use in a national training course "Endovascular 101: Introduction to Cerebral Angiography."

This study demonstrates a close collaborative relationship between the design research team and the clinical team. The lack of ergonomic knowledge in catheter-based procedures as indicated in ANSI/AAMI HE 75 proved a fertile area for exploration (AAMI, 2009). This CI study required an IRB approval and was conducted within an academic institution. Key takeaways from this case include close collaboration can improve the ability to have advanced set-up in the clinical space, provide ongoing feedback that insights and interpretations generated are truly valid and the information produced as a result of the study goes beyond meeting the needs of medical device development teams and into clinical training.

BIOMARKER STRATEGIES SNAPPATH® CASE STUDY: CI STUDY TO INTEGRATE A BREAKTHROUGH DIAGNOSTIC SYSTEM INTO THE CLINICAL ENVIRONMENT *TOR ALDEN*

Chapter 9 addresses a process to uncover the development needs of a first-of-its-kind automated live-cell processing system designed to enable highly predictive tests for targeted drug therapies. BioMarker Strategies developed the breakthrough technology to allow biomarkers to be interrogated in live cells outside the human body (*ex vivo*) using fine needle aspiration or other types of biopsies.

The SnapPath System required an appropriate design for multiple settings and markets, including preclinical academic research, companion diagnostics, and (Bio) Pharmaceutical drug development. Key takeaways from this case include the process used to interview key users in order to understand the environment and workflow that lead to successful user acceptance. This process was aimed at transitioning a highly technical scientific program into commercialization with focused innovation into established protocols. This required the research and development team to determine the key elements in the design, as defined by stakeholder. Then the findings were translated into system architecture and system workflows for continued product development. Their successful efforts produced a Silver award in the Medical Design Excellence Awards.

This case is unique as it represents CI study methodology used in a mixed hospital/laboratory environment with multiple users and user groups rather than a specific in-hospital device. The development team worked in close collaboration with users to identify key design considerations. During the process key assumptions were challenged, which shifted the overall development program in a positive direction.

CONTEXTUAL INQUIRY AS A TOOL FOR MEDICAL-DEVICE DEVELOPMENT: THE CASE OF HARMONIC FOCUS
STEPHEN B. WILCOX, PHD, FIDSA

The case study in Chapter 10 is about using contextual inquiry to change the product-development process at a major corporation and to change the role of industrial design within that process.

It is based on the study of a harmonic device that is a type of energy instrument that uses high-frequency vibration instead of an electrical current to transect tissue hemostatically, that is, to simultaneously cut and coagulate. The goal of the program was to provide an understanding of thyroidectomy procedures for the team that was charged with developing a new harmonic device for those procedures.

This case study walks the reader through the steps taken from kick-off, planning, sample size determination, recruiting, protocol development, informed consent, education of the research team, and then the actual conduction of the research. Further discussion demonstrates an information pyramid from the least-refined data to organized detailed spreadsheets. In this case, data visualization consisted of a multimedia report, posters in the form of a procedure map.

This project extended over a several-months-long development process that involved a number of rounds of design/prototyping and usability testing (among many other things) to work out all the details and assure that the device was optimized. The subsequent device design won a Gold Industrial Design Excellence Award for the project.

This successful program describes close collaboration from a research consultant and a major corporation. The impact of the study affected not only the product being designed but also internal practices. Key takeaways from this case include an additional example of process, various uses for the results of a CI study, and overall positive organizational impact.

USING CI TO INFORM DESIGN DEVELOPMENT OF AN INCISION AND DRAIN PACKING TOOL FOR USE IN EMERGENCY MEDICINE
MARY BETH PRIVITERA

The goal of the CI study found in Chapter 11 was to capture the entire clinical procedure of incision and drainage (IND) for the treatment of skin abscesses. An indepth analysis of the procedure from the time of patient presentation to clinical decision-making, treatment, and subsequent gauze removal is included. Additionally, a detailed description of the equipment used to prepare and complete IND treatment and the exploration of clinician perspectives were analyzed. The contraindications and variances of procedure across a large clinical team were determined through the CI process. This information was then translated into a device development opportunity.

This unique case began with communicated problem, then was fully investigated by a multidisciplinary team with full collaboration and participation of the entire Department of Emergency Medicine at the University of Cincinnati. The process included clinical immersion, extensive interviewing with clinical collaborators, and exploration in simulation model development as well as iterative design development.

The study presents challenges faced by the research team and how they were mitigated. The process, findings, and design development are detailed within this case. This successful program generated several functional prototypes and intellectual property for the UC.

REFERENCE

AAMI, 2009. ANSI/AAMI HE75, 2009/(R)2013 Human Factors Engineering—Design of Medical Devices, USA.

CHAPTER 8

A CI Study in Cerebral Angiography for Product Development Strategy and Clinical Training

Mary Beth Privitera* **and Andrew Ringer†**
*University of Cincinnati and Know Why Design, LLC, Cincinnati, Ohio, USA
†University of Cincinnati and Mayfield Clinic, Cincinnati, Ohio, USA

Contents

8.1 BACKGROUND

A cerebral diagnostic angiogram is an interventional radiology procedure performed to assess potential cerebrovascular health or blood flow in the brain. During the procedure, a catheter (thin, long tube) is inserted through the femoral artery at the groin and then navigated through the patient's vascular system. This includes traversing the aortic arch, carotid arteries, and ultimately into the brain.

Guidewires (smaller gauge wires) are inserted within the catheter to aid in vascular navigation. Dye, known as contrast, will be injected into the catheter to generate images under fluoroscopy (X-ray) and clinically assess location of the catheter and the artery(s) of interest. This procedure

M.B. Privitera: Contextual Inquiry for Medical Device Design.
DOI: http://dx.doi.org/10.1016/B978-0-12-801852-1.00008-3

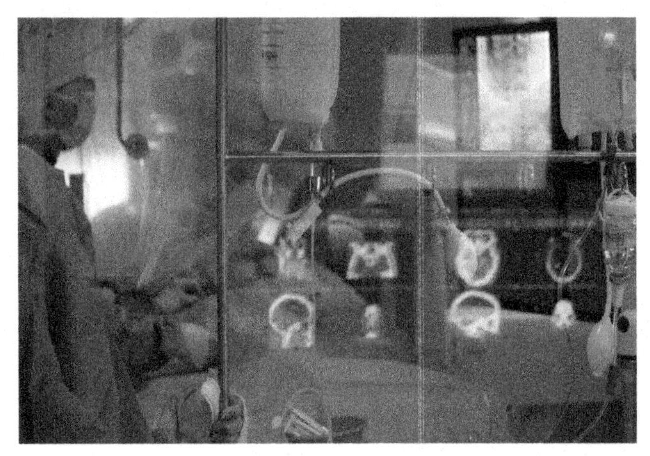

Figure 8.1 View of the operatory during a cerebral diagnostic angiogram.

is used to diagnose diseases like vascular malformation and intracranial aneurysms.

The procedure is conducted in an interventional radiology suite that consists of a monitor bank for reference during the procedure, a fluoroscopy unit and patient table along with a sterile equipment table. Additionally, there is a control room for monitoring the procedure (Figure 8.1).

The goal of this program was to work collaboratively with the Department of Neurosurgery in order to identify key behaviors and actions which attributed to challenges in the procedure, identify new product development opportunities, and lastly to develop training materials for use in a national training course "Endovascular 101: Introduction to Cerebral Angiography." This research was sponsored by NSF Grant #IIP-0652208 and industry members of the NSF MIMTeC I/U CRC.

8.2 METHODOLOGY

This study went through considerable planning and institutional review board (IRB) approval was necessary due to the intention of publishing results as required by the NSF Industry/University Center sponsorship. To start the process, the clinicians provided the research team a series of training manuals and explained the procedure in conference. An additional literature review was conducted by the team prior to beginning the data collection series that included 97 articles on ergonomics as applied in catheter device use, development, and the procedure of cerebral diagnostic angiograms.

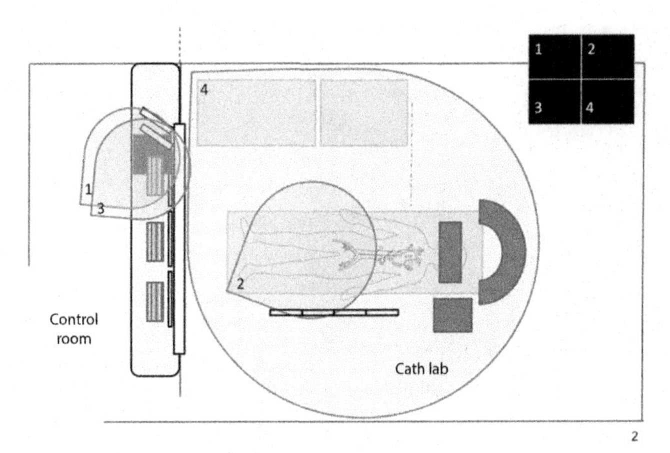

Figure 8.2 Overview of video camera placement.

In order to record the procedure and capture the overall environment, the hands of the surgeon and the fluoroscopy image, a system of photography, and videography units were strategically placed within the surgical theater (as shown in Figure 8.2; numerical figures represent data recording locations).

Although we sought approval for 50 cases, a total of 24 diagnostic cerebral angiogram cases were found to be sufficient data in order to compose insights. Clinician collaborators, Todd A. Abruzzo, M.D. and/or Andrew J. Ringer, M.D., guided each case. In some cases instruction was provided to one of five fellows (trainees) with a ranging degree of experience—first year fellowship experience to third year fellowship experience who were also being observed. Each diagnostic cerebral angiogram case required the supervision of at least two members of the research team to control the multichannel video recording system and collect notes on the proceedings of each case. The research team was scheduled to arrive prior to the procedure in order to assure consent was granted and then set-up the video equipment prior to the arrival of the patient into the angio suite. Throughout data collection both Drs. Abruzzo and Ringer were interviewed in order to confirm findings and assure correct translation for the purposes of design.

8.3 DATA ANALYSIS AND FINDINGS

Data analysis was planned prior to the start of the study. As video was collected it was prepared for analysis in our multichannel system

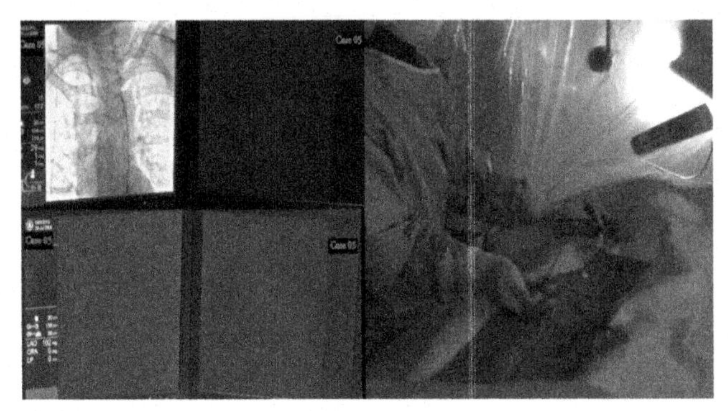

Figure 8.3 Data imported into our multichannel system for analysis.

(Figure 8.3).Using the system the research team was able to synchronize the video thus capturing the hand movements, the catheter movements as shown on the fluoroscopy monitors, and the what was happening in the room overall.

Notes were also taken throughout the procedure. Notes consisted of personnel in the room: attending/s, fellows, technologists, nurse, and equipment specialists. Their responsibilities were listed as the procedure progressed. In addition, detailed equipment lists and medication preparations (examples below) were taken at every procedure for comparison. This enabled the research team to understand device preferences and the reasons why.

Equipment List

1—Applicator Chloraprep (10.5 mL)
2—Bag Band (36" × 28", clear with tape)
1—Bowl Guidewire (2500 cc with lip)
1—Bowl Solution (32 oz. 1000 cc)
4—Clamp Towel (4-1/2 Plastic Blue)
1—Cover BK Table (44" × 78", ZN RNF)
1—Cup Denture (8 oz., Turq. with lid)
1—Cup Medium Grad (2 oz. 60 cc)
1—Cup Medium Grad (2 oz. 60 cc, Yellow)
1—Forceps Mosquito (5" Halsted Straight)
2—Gown Surgical XL (UNRG SMS)
1—Holder Needle 6 Crille Wood
1—Insert—PHS LG Contents
1—Kit CSTM—Health Alliance
2—Label (3) for Skin Marker
1—Label Med CSTM—Health Alliance
1—Lidocaine HCL 1% 30 mL Vial G

1—Marker Utility Reg Tip Wide Body
1—NDL Reg Bevel 18 G × 1.5" TW PI
1—NDL Reg Bevel 25 G × 1.5" Blue
1—Pocket Foam 50 cc with Flap
1—Pouch (19" × 27")
1—Scalpel Surg #11 SS
1—SCS O.R. 5.5" SH/SH STRT
40—Sponge Gauze (4" × 4" 16 Ply)
1—SYR 10 cc L/L YLW PLNGR
2—SYR 10 cc LL
3—SYR 20 cc LL
2—Tape Back Table Cover
6—Towel O.R. (17" × 26") Blue LF
1—Tray Foam (14" × 18")
1—Workhorse Catheter 5F Tapered Glide Catheter*
1—Guidewire (type of stiffness is dependent on patient anatomy)*
0.035" Angled Glidewire (150 cm) or 0.038" Bentson Cerebral Guidewire (145 cm)
1—Introducer Set—Micropuncture Introducer Silhouette Transitionless
21 G (7 cm) Entry Needle
0.018" (40 cm) Guidewire
5F (10 cm) Sheath
1—Stopcock (Morse Stopcock 1050 psi)
1—Medium Power Injector
1—GE Innova 4100 Biplane Imaging System
1—GE LOGIQ i Ultrasound System
1—Interventional Radiology Imaging Processing Software

Medical Preparations and Solutions

15—Heparin Saline Syringes (15 mL each)
1—*x* mL of Visapaque 270 or Isovue 300 Contrast Medium
N—Versed (Conscious sedative)
N—Aspirin (Anticoagulant via platelet inhibition)
N—Heparin (Anticoagulant)
N—Ancef (Antibiotic)
N—Fentanyl (Narcotic Analgesic)
N—Benadryl (Antihistamine)
N—Morphine (Narcotic Analgesic)

* Equipment denoted as variable case to case dependent on surgeon preference and patient anatomy.

Contrast Agent

Visapaque 270 (Iotrolan) or Isovue 300 (Iopamidol)

- Selection criteria is based on osmolarity, ionic charge, expense, and patient contraindications
  - Contraindications include arise from hyperosmolarity, which may induce pain during injection, cardiac overload, and renal toxicity
  - *Recommendation*: Use nonionic contrast, which has a lower incidence of common systemic complications, however, is more expensive. Patients with normal and healthy renal and cardiac functions can typically tolerate several hundred milliliters of contrast without signs of complications
- *Standard contrast agent*: Iodine (highly absorbing to X-ray exposure)
  - Contrast Agent: Hyperosmolar—320–1700 mOsm (blood is approximately 300 mOsm)
- Method to Limit Contrast Load per Procedure:
  - Monitor total usage throughout procedure
  - Create a plan of how much contrast will be needed according to imaging sequence plan
  - Dilute contrast when possible
  - For catheter placement confirmations, only use a small amount of contrast (1–3 mL) during injection and imaging sequence
  - Verify if catheter end-effector is placed properly, so that contrast bolus efficiently flows through proper arterial branches

A task analysis was performed in order to identify each step in the procedure and begin the process of challenge identification and the clinical decision-making process. Examples for each step are provided below.

Patient and Room Preparation

1. Bring patient into imaging suite using surgical bed.
2. Transfer patient to translatable bed of Innova Biplane Imaging System and place in supine position
3. Place intravenous line into patient's antecubital fossa (cephalic, basilica, or median cubital).
4. Place blood pressure cuff on patient's right or left arm (dependent on needs to pressurize patient's right or left cerebral venous system).

5. Inject 150 mL of Versed through IV access to patient for conscious sedation (monitor and inject throughout procedure as condition warrants)
6. Use Chloroprep applicator to sterilize patients right or left femoral region of access
7. Move and position GE Innova 4100 Biplane Imaging System C-Arm around cephalic region of patient
8. Move (8) screen system to lateral and distal most location with respect to patient
9. Place surgical drape to cover chest to feet area around femoral insertion location
10. Connect tubing lines to Power Injector
11. Place (3) bags of solution on rack near feet of patient
12. Cover transparent protection screens with plastic
13. Position transparent protection screens over patient to provide protection to patient and surgeons

Locating Femoral Artery

1. Use a hemostat and fluoroscopy to locate femoral head in patient's leg (location is selected on previous insertion location, potential for atherosclerosis, and size of vessel)
2. Mark the skin with a marker to represent femoral artery location
3. Inject 1% Lidocaine into skin and subcutaneous tissue to create wheel in region of femoral artery insertion
4. Use ultrasound to locate femoral artery and assess atherosclerosis of vessel, then mark at distal and proximal location of artery
5. Create a 1–2 cm skin incision using an 11 blade to expose common femoral artery
6. Use mosquito clamps to expand the incision site
7. Use ultrasound guidance to insert 21 G entry needle at a 45° angle through femoral artery
8. Puncture femoral artery using single wall or double wall puncture technique
9. Use the nondominate hand to hold the needle and secure back-bleeding from the hub when back-bleeding is achieved
10. Pass the 0.018" (40 cm) guidewire through the entry needle
11. Remove the entry needle leaving the 0.018" (40 cm) guidewire in place

12. Place a 5F (10 cm) sheath over the 0.018" (40 cm) guidewire and into the femoral artery
13. Remove the 0.018" (40 cm) guidewire leaving the 5F (10 cm) sheath in place
14. Flush the 5F (10 cm) sheath manually with heparanized saline
15. Insert 0.038" (145 cm) Cerebral Bentson guidewire through 5F (10 cm) sheath
16. Remove the 5F (10 cm) sheath and replace with a longer 5F sheath containing a side port attachment for heparin/saline solution connection
17. Remove 0.038" (145 cm) Bentson Cerebral guidewire
18. Manually flush longer 5F sheath with heparin/saline solution
19. Insert 0.035" (150 cm) angled guidewire through 5F (100 cm) Angled Glide Catheter
20. Flush 0.035" (150 cm) angled guidewire through 5F (100 cm) Angled Glide Catheter with heparin/saline solution
21. Using fluoroscopic guidance, insert 0.035" (150 cm) angled guidewire and 5F (100 cm) Angled Glide Catheter assembly through longer 5F sheath

Catheter and Guidewire Advancement

1. Advance 0.035" (150 cm) angled guidewire and 5F (100 cm) Angled Glide Catheter assembly with guidewire just inside distal tip of catheter to aortic arch by passing through the following arterial structures all while under fluoroscopic guidance:
   - Femoral artery
   - Common iliac artery
   - Inferior mesenteric artery
   - Superior mesenteric artery
   - Descending aorta
   - Aortic arch
   - Brachiocephalic (innominate) artery
   - Right common carotid artery
   - Vertebral artery
   - Surgical site of interest

   Note: Use preferred torque technique rather than a push technique to advance guidewire, which may prevent unnecessary dissections.

Guidewire Device Exchange

1. During advancement, assess device type, size, and end-effector shape and determine if device should be exchanged for a different configuration for proper advancement through tortuous anatomy.

2. If removing or replacing a guidewire:

a. Prepare replacement guidewire by wiping and/or soaking with saline and if needed manually manipulate distal tip of guidewire for optimal shape.

b. Remove guidewire from catheter by using the nondominate hand to firmly grasp the catheter and the dominate hand to remove the guidewire. Catheter should remain in place within arterial system.

Note: Use fluoroscopy to monitor distal end of catheter to verify no abrupt movements during guidewire removal.

c. While removing guidewire, wipe down guidewire with saline-soaked gauze sponge in the direction toward the patient (moving toward patient prevents catheter from being pulled out of artery).

d. Spool guidewire into packaging loop and then place into saline bath on sterile table.

e. Connect saline syringe to stopcock on end of catheter hub to draw back blood in the catheter and remove "dead space."

Note: Catheter lumen should always contain saline or a guidewire to prevent clotting of blood.

f. Tap a second clean saline syringe with finger to migrate air bubbles to exit point and then plunge out until a droplet is created at exit point.

g. Connect clean saline syringe to stopcock on end of catheter hub by making a wet-to-wet connection in a vertical orientation.

Note: Air bubbles should be prevented at all costs by using syringe to remove air bubbles and making connections in a vertical and wet-to-wet configuration.

h. Inject x mL of saline using syringe in vertical orientation (90°) to replace any new blood in the catheter with saline.

Note: Never fully evacuate saline syringe of contents to prevent injection of air into arterial system.

i. Place thumb of nondominate hand over the opening of stopcock or close stopcock to equilibrate system and prevent leakage of saline from catheter lumen.

j. Under fluoroscopy guidance, insert replacement guidewire of choice into catheter by using nondominate hand to grasp catheter and using dominate hand to slowly advance guidewire from packaging loop to catheter.

k. Advance replacement guidewire just inside distal most tip of catheter.

Note: Guidewire should always be placed just inside the distal most tip of catheter when not in use to prevent vessel dissection and trauma.

l. Aspirate and flush catheter line with heparin/saline solution.

m. Inject contrast to confirm placement and ensure catheter tip is in proper location for power injection and/or imaging.

Catheter Device Exchange

1. During advancement, assess device type, size, and end-effector shape and determine if device should be exchanged for a different configuration for proper advancement through tortuous anatomy.
2. If removing or replacing a catheter:
   - **a.** Prepare replacement catheter by soaking and/or wiping down with saline.
   - **b.** Verify replacement catheter is free of air bubbles by wiping down with gauze sponge.
   - **c.** Place pinky and ring finger of nondominate hand to hold guidewire in place.
   - **d.** Remove catheter by placing forefinger and thumb of dominate hand to gently remove catheter out of arterial passage.

     Note: Use fluoroscopy to monitor distal end of guidewire to verify no abrupt movements during catheter removal.
   - **e.** Flush and wipe down removed catheter with saline solution.
   - **f.** Place removed catheter in saline solution bowl.
   - **g.** Slide replacement catheter over guidewire and advance toward incision.
   - **h.** Pass catheter through skin and tissue over the stiff portion of guidewire.
   - **i.** Incrementally advance catheter over guidewire through arterial system a few centimeters at a time under fluoroscopy guidance.

     Note: A second surgeon aid can provide assistance by grasping guidewire firmly to stabilize guidewire during advancement.

Contrast Imaging Sequence

Manual Injection

1. Tap syringe filled with contrast while in a vertical orientation to remove air bubbles.
2. Attach a syringe filled with contrast to stopcock at hub of catheter using a wet-to-wet and vertical orientation.
3. Instruct patient to not breathe, move, or swallow during imaging sequence.

4. Inject a small amount of contrast while activating imaging sequence to check for appropriate catheter location within clinical region of interest.
   a. *Note:* Do not allow syringe plunger to deplunge (deplunging will allow air into system), inject using palm of hand not thumb, inject using one smooth accelerated injection, and do not use entire contrast which may inject air into system.
   b. *Equipment*: Smaller the syringe, the higher the positive pressure levels during injection.
   c. *Recommendation*: Not recommended for 3D image sequences due to the amount of contrast needed.
   d. *Warning*: Increased radiation exposure occurs during imaging sequence.
5. View image sequence through video and/or still shots to confirm catheter placement.
6. Adjust catheter if need be and inject additional contrast to confirm catheter placement.

Power Injection

1. Connect tubing lines to Power Injector using luer-lock connections.
2. Verify all air bubbles have been removed from line.
3. If air bubbles are present, connect syringe and evacuate line and disconnect syringe.
4. Aspirate catheter through injector by turning knob on Power Injector until blood returns into line.
5. Instruct RT for proper injection settings (amount of contrast, flow rate, and time delay).
6. Instruct patient to not breathe, move, or swallow during imaging sequence.
7. Instruct all personnel in suite to move into control room.
8. Instruct RT to begin power injection imaging sequence.
9. View image sequence through video and/or still shots to confirm catheter placement.
   a. *Note*: For 5F (65–100 cm) catheter, use higher pressure (1050 psi) to produce adequate bolus of contrast.
   b. Recommendations for use: Good source for constant rate and pressure of injection, efficient for viewing anatomical structures requiring high volumes of contrast (e.g., aortic arch).
10. Adjust catheter if need be and inject additional contrast to confirm catheter placement.

Clinical Assessment

1. View imaging sequence and evaluate clinical therapy if needed immediately or plan clinical action for secondary procedure.

Procedure Wrap-up

1. Gently remove catheter until the catheter end-effector reaches the common iliac artery.
2. Under fluoroscopic guidance, continue removing the catheter through the femoral artery just prior to the insertion site.
3. Check insertion site for trauma (e.g., dissections) by injecting a small amount of contrast and viewing imaging under fluoroscopy.
4. Verify imaging shows no dissections within femoral artery.
5. Seal the insertion site by either applying pressure or using a closure device:

   Applying Pressure

   a. Slowly remove the catheter until it is free from patient's arterial system.
   b. Apply pressure to insertion site with gauze to maintain the arteriotomy.
   c. Slowly remove the sheath and again apply pressure on insertion site to maintain arteriotomy.

      Note: For 4F or 5F catheters, pressure should be applied for 15 minutes to allow arteriotomy to clot and maintain hemostasis; for larger size catheters or if heparin is used, pressure should be applied for longer than 15 minutes to allow arteriotomy to clot and maintain hemostasis; anticoagulant medication dosage will influence rate of hemostasis.

   Closure Device

   a. Select AngioSeal or Perclose closure device (or similar).
   b. Remove the sheath from the insertion site.
   c. Slide the closure device sheath over catheter.
   d. Remove catheter once closure device sheath is in place.
   e. Consult device-specific instructions for use for further closing requirements.

8.4 DESIGNING THE INFORMATION FOR ANALYSIS AND COMMUNICATION

Many times during interview the attending physicians would draw the situation they were trying to avoid or to further explain the behaviors of

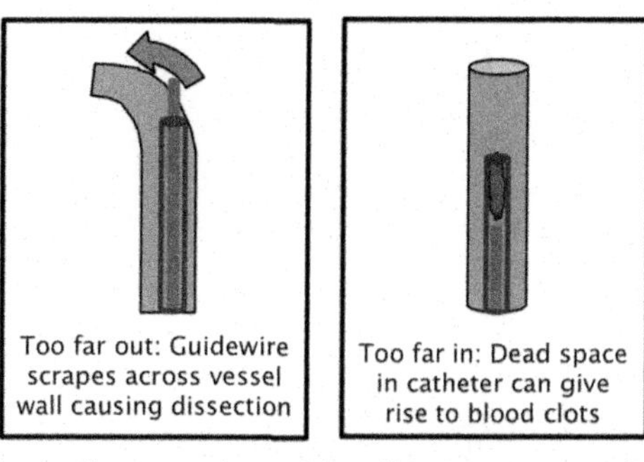

Too far out: Guidewire scrapes across vessel wall causing dissection
Too far in: Dead space in catheter can give rise to blood clots

Figure 8.4 Guidewire, catheter, and arterial wall interactions at the distal tip.

the catheter based on their experiences (Figure 8.4). For example, gaining a clear understanding of the interaction between the amount of guidewire extension relative to the guide catheter while inserted into the artery was explained as a 3 step process described below and the clinical implications were illustrated.

3 Step Process:

I. Guidewire stays inside the tip of the catheter until ready for use
- Increased risks associated with being too far in or out:
- Too far in: Blood clots due to large amount of dead space in the catheter
- Too far out: Dissection of the blood vessel

II. Guidewire advances distally into the blood vessel
- Distal guidewire access offers anchored stability for catheter advancement

III. Catheter advances along the guidewire until it returns to position I.
- Torque is applied to the catheter to:
- Aid in following guidewire contours
- Reduce friction

Clinical Implications:

Advancing the guidewire before the catheter offers less trauma to the vessels and makes arterial subselection easier.

Other techniques were also explored using visualization techniques. For example, the video data recorded on the fluoroscopy was illustrated in order to communicate precisely how the guidewire interacts within the anatomy in both positive manners (Figure 8.5) and negative situations (Figure 8.6).

In Figure 8.6, catheter redundancy is illustrated. This is a key challenge for users and is related to the proximal input $\neq$ distal response.

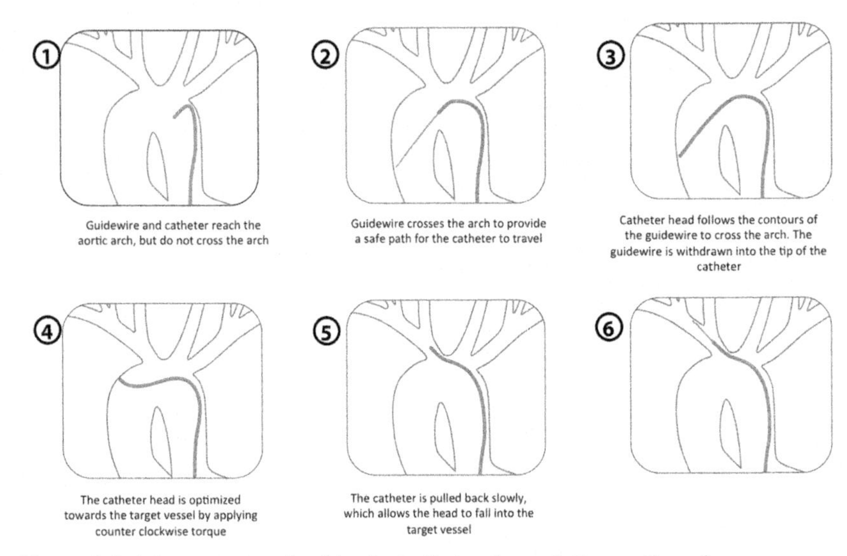

Figure 8.5 Advancement of guidewire/catheter through the aortic arch.

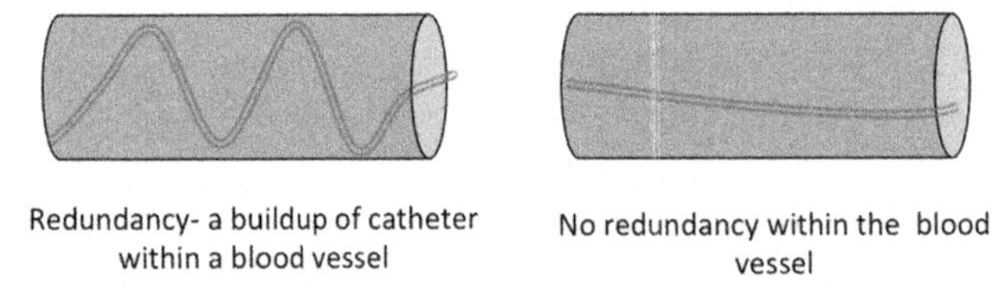

Figure 8.6 Catheter redundancy illustrated generically.

The amount of loss in distal response can be attributed to a number of factors:

- Stiffness of the catheter—Braided versus non-braided

  Braided catheters are more rigid than non-braided catheters and therefore offer more control
- How distally located the catheter is

  More distal location means greater contact forces and therefore more resistance to motion
- Redundancy in the system

Greater redundancy increases friction and creates mechanical losses in distal response due to poor line of action. It also builds up elastic potential energy in the catheter, when this energy is released, it can cause the catheter to make sudden, traumatic movements resulting in loss of positioning, perforation, or dissection.

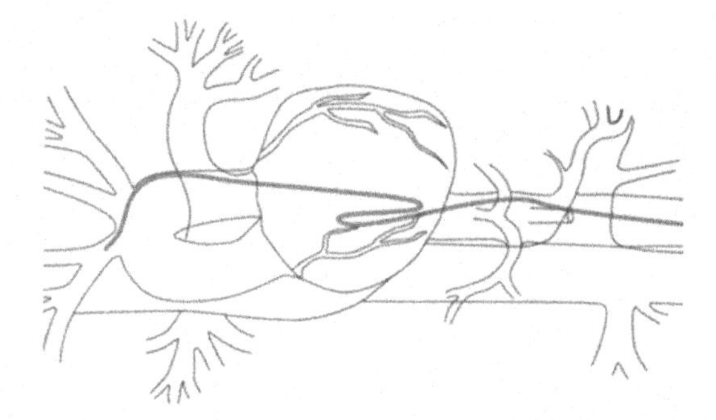

Figure 8.7 Redundancy in catheter location 1.

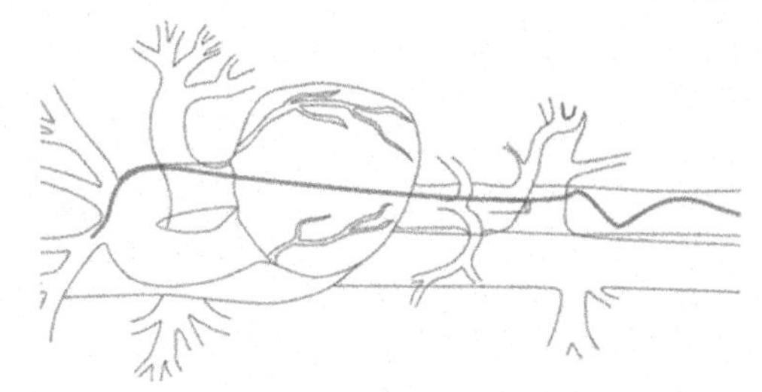

Figure 8.8 Redundancy in catheter location 2.

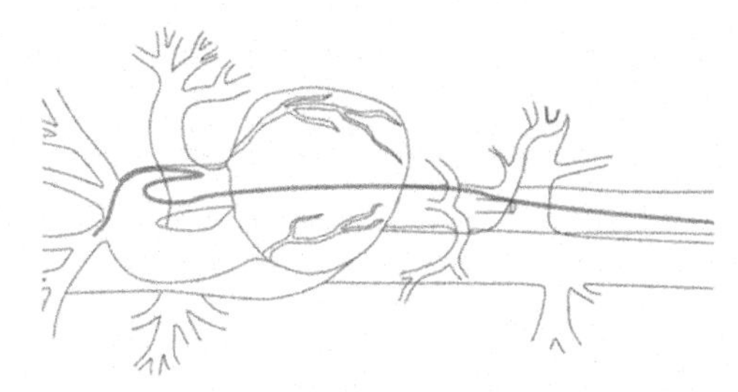

Figure 8.9 Redundancy in catheter location 3.

The clinical implication is these factors make predicting product response consistently difficult for those lacking extensive experience. Redundancy can happen at any time during the procedure and typically happens in the aorta (Figures 8.7–8.9).

While the above challenges were all explicitly communicated during interview and observed in the cases, the research team also noted implicit challenges. For example, bubble mitigation. In analysis of time spent

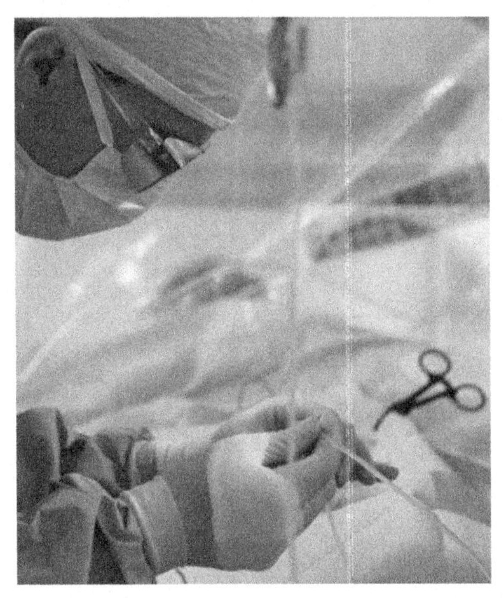

Figure 8.10 Inspection for bubbles 1.

mitigating bubbles it was noted that the average time spent assuring that there were no bubbles in the line was 30 minutes out of a total average procedure time of 90 minutes. Bubbles in the line can lead to stroke or other complications. It is imperative that the clinical team be diligent in this task for patient safety reasons. The means of removing the bubbles consisted of close inspection, flicking the line, or using a hemostat to flick the line that sends the bubbles distal (Figures 8.10 and 8.11).

Once analysis was complete the data was condensed into the most important elements and roughly mapped without images for physician review (Figure 8.12). Several copies of the map were provided to each physician involved in the study. This assured agreement on priority. Once the reviews were complete and text edits were made, the entire map was drafted and printed with inclusion of illustrations and images (Figure 8.13).

8.5 DETERMINING PRODUCT DEVELOPMENT OPPORTUNITIES

Once the map was complete, the collaborative team was able to take a step back and identify areas of opportunity for design. As a result, there were two challenges of particular interest to the design team—

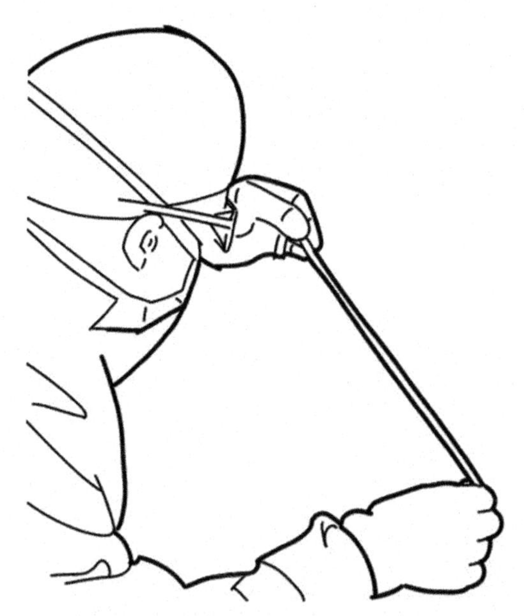

Figure 8.11 Inspection for bubbles 2.

redundancy and bubble removal. The insights provided in the map resulted in the following need statements:

1. To design a catheter user interface design to prevent redundancy and provide better tactile feedback and
2. To design an assistive tool to facilitate removal of bubbles.

These problems were then given to multidisciplinary graduate students in the Medical Device Innovation & Entrepreneurship Program (MDIEP). Note these students had no involvement in the contextual inquiry (CI) study. They used the procedure map coupled with a minimal amount of observations in order to conceptualize potential solutions.

Team one took on the user interface of the catheter and the redundancy problem. Figure 8.14 shows conceptual exploration of a fundamental change in catheter geometry targeted solely at the location of interaction with the user. It is based on the theory that perceptively it is difficult to know the amount of rotations produced between the thumb and index finger when rotating a tube. In changing the geometry, users would be able to have a marker in an edge surface that may be able to provide feedback. Subsequently, one of these concepts was selected for advanced modeling and engineering analysis.

Step	1 Set-up	2 Access	3 Advance to aortic arch	4 Advance to inominant art	5 Prep for Angiogram	6 Injection: Manual	6 Injection: Power	7 Angio Review	8 Leg Run: Close
Users (Primary / Secondary)									
Task	Prep and drape Set-up all tools anticipated Organize work table Fill & label syringes Line set-up: Sheath connected to flush with syringe connection Flush lines: remove air	Identify landmarks: iliac crest pubic symphosis 3 fingers below Make small incision Microneedle puncture Microwire exchange inserted Insert wire for sheath Exchange dilator for sheath Suture in place	Advance catheter over guidewire. Watch wire from groin to arch Position wire just beyond inominent artery Advance cath along wire	Pull wire back appox. 1" Using both hands-Twist assembly towards operator (lft on catheter- rt on hub) Position so that internal cath points up (towards patient head) Pull catheter back to engage into inominant artery Position flouro	Advance catheter over wire into r. subclavian follow to target artery Remove wire, wipe w/ gauze Turn off flush, if connected Aspirate Flush catheter with saline Check catheter position	Check contrast syringe for bubbles - remove if found Attach syringe using wet to wet technique (vertical) Instruct patient not to "breathe, move, or swallow" In sequence, activate image depress foot pedal-inject contrast	Verify all lines are bubble free Connect lines Aspirate cath through inj by turning knob till blood is visible in line Instruct RT for proper settings (contrast, flow, & time delay) Instruct patient "don't breathe, move or swallow" All personnel leave room Instruct RT to begin imaging	Initial image review: proper position of cath proper artery target Evaluate clinical condition/s Plan clinical action for secondary procedure	Gently remove cath till CIA Image groin Advance Angioseal wire in sheath Remove sheath over wire Thread Angioseal device Deploy suture Check for hemostasis Check distal pulse Apply dressing
Tools	5F Tapered Glide Catheter Guidewire 5F Microneedle Microwire exchange Microdilator Birdbath Syringes: contrast AngioSeal 3-way or TBV	Scalpel Microneedle Microwire Exchange dilator Sheath Suture Torque device	Sheath Angled Glidewire (150cm) or Bentson Cerebral Guidewire (145cm) 5F tapered Glide Catheter	Sheath Angled Glidewire (150cm) or Bentson Cerebral Guidewire (145cm) 5F tapered Glide Catheter	Sheath Angled Glidewire (150cm) or Bentson Cerebral Guidewire (145cm) 5F tapered Glide Catheter Syringe 4x4 Gauze	Sheath Angled Glidewire (150cm) or Bentson Cerebral Guidewire (145cm) 5F tapered Glide Catheter Syringe 4x4 Gauze	Sheath Angled Glidewire (150cm) or Bentson Cerebral Guidewire (145cm) 5F tapered Glide Catheter Syringe 4x4 Gauze	Monitor	Sheath Angled Glidewire (150cm) or Bentson Cerebral Guidewire (145cm) 5F tapered Glide Catheter Syringe 4x4 Gauze
Challenges	Removing air from line/s Recognizing air in line/s	Too large of incision Too much force on puncture Wire over inserted Wire inserted to unindended location	Redundancy in system ** see illustrated examples** inadvertent arterial damage	Redundancy in system ** see illustrated examples** inadvertent arterial damage	Removing air from line/s Recognizing air in line/s	Removing air from line/s Recognizing air in line/s	Catheter can slip out of vessel by the force of bolus injection For power injection: RT's may not ***handle/lever that resets image planes??****	Assuring imaging correct vessel??	Pulling cath out too fast
Mitigation	Flush saline Flick with fingers, tools Inspect	Suture Image Start access step again Ask for help	Vigilance of hand movement to cath response monitored	Vigilance of hand move-ment- cath resposne Pull back on cath to straighten	Flush saline Flick with fingers, tools Inspect	Flush saline Flick with fingers, tools Inspect	Flush saline Flick with fingers, tools Inspect	Puff injection verify placement Monitor contrast dosage throughout procedure Plan contrast use according to image sequence plan Verify cath placement so that contrast bolus efficiently flows through proper arterial branches	Vigilance in knowledge of cath location to relative anatomy
Insights	Users have preferrential tools derived from training In general, the risk is proportional to the challenge: If it seems too difficult choose another method to obtain same information	Once mastered, this step becomes fast and routine w/high degree of predictability Can be a challenge for patients w/weak pulse Take time this requires caution	Interplay between the wire and the cath must be managed Be efficient: avoid wasting time getting up the arch	Cath position/direction towards target requires vigilance in proprioception Elongated arch = increased angle access as well as increased risk Navigating to target requires finesse: avoid cutting corners	Mastery of connections: wet to wet vertical approach Plan contrast usage accord-ing to image sequence Goal: best possible image w/least amount of radiation Check for bubbles	Image quality is directly related to timing skills of the user	More vascular anatomy can be seen with this method but there is an increased force of contrast delivery Force of injection can cause cath to slip from position	Overall efficiency of procedure can be compromised Increased contrast /radiation dosage can result Before moving on make sense of the image	
Variance	Set-up use preference: 3-way stop cock Touhy Borst	Use of lidocaine: Sedate vs. Non-sed patient Use of Glidecath Use of Bentson wire Use of torque device	Expect anatomical variability: elongated aorta, age, etc. which will factor into approach	Ideal Progression: r. vert, r. common, lft. common, lft.vert	2-way method: Aspirate under floro: inject small amounts of dilute contrast (3:1) to verify cath position Locks 2-way - connect flush Toux method: Catheter connected to flush 30 cc/hr regulator			Contrast Agent: Visapaque 270 (Iotrolan) Isovue 300 (Iopamidol) Hyperosmolar 320-1700 mOsm	Presure Method: apply pressure to insertion site with gauze, slowly remove sheath then resume pressure (approx. 15 mins. for 4F/5F cath, longer if heparin is used)

Figure 8.12 Draft copy of clinical map text for physician review.

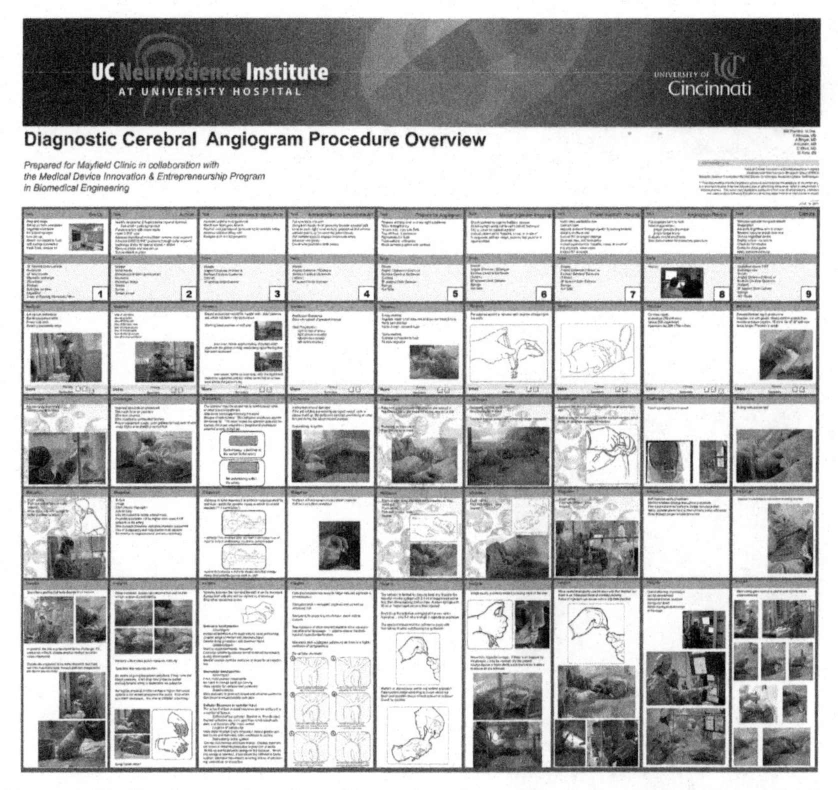

Figure 8.13 Final procedure map illustrating the steps of diagnostic cerebral angiogram procedures.

Figure 8.14 Conceptual catheter cross-sections as designed by the MDIEP team.

Cecelia Arredondo, Master of Design student, explored the bubble identification and removal challenge in full detail based on limited engineering support from MDIEP engineers. Her process included full conceptual design, functional modeling (Figure 8.15), and testing with users. The premise of the device was to use light, magnification and vibration as the key to bubble management.

Figure 8.15 Functional prototypes of bubble mitigating tool.

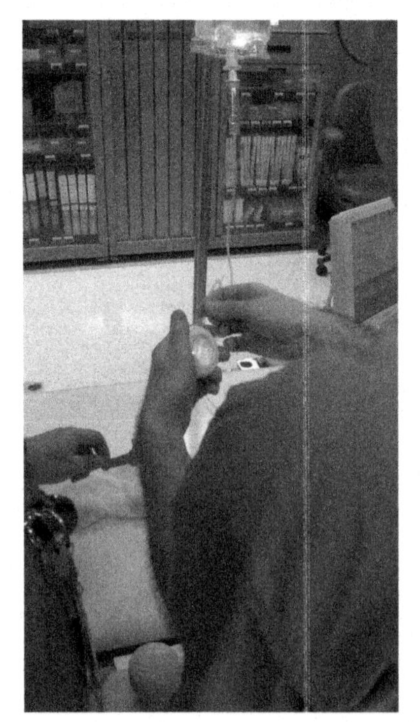

Figure 8.16 Testing functional models of the bubbler device in the angio suite with users.

This CI study yielded robust problem definition and opportunity for product development. Further, by using the results, a legacy of procedural information was quickly and easily communicated to design teams not involved in original study. Lastly, the collaborative partners of physicians, designers, and engineers are still working together today (Figure 8.16).

8.6 CLINICAL INVOLVEMENT AND THE DEVELOPMENT OF SEMINAR-FRIENDLY TRAINING MANUALS/INTERACTIVE TOOL

This study was proposed by a group of biomedical engineers and admittedly met with some skepticism on the part of the clinician partners. Medical procedures are typically performed through a series of routines developed organically and any disruption of the routine, real, or perceived is typically unwelcome. The prospect that the routine could be improved through systematic observation and analysis was foreign to the clinician partners. Despite their trepidation, the process proved painless, as the observation team was entirely unobtrusive and provided feedback only after eliminating less useful information. The resulting relationship proved very productive, spawning several projects from device design to procedural instruction and process improvement.

The insight the clinicians gained from the in-depth analysis of a common procedure allowed the clinicians to identify rate-limiting steps in the procedure, sources of potential error, and opportunities to increase efficiency. While the large format of the procedure map proved difficult to use in one-on-one teaching situations, it has proven valuable in larger group settings and became an important component in the national course conducted by our clinical partners (Figures 8.17–8.25).

8.7 THE CHALLENGES AND HOW THEY WERE MET

Initial response to the concept of placing multiple cameras in the Angio suite and watching clinical care closely was initially met with apprehension. While one senior clinical faculty member promoted the study, the other was hesitant. After a bit of trial and error in the suite without patients, the research team determined the optimum locations of the cameras and the means by which they would be fixated during the actual clinical cases. As described by the clinical team, the research team was unobtrusive and soon the clinical team realized that the discussions with the research team were actually mutually beneficial. The original intent of the study did not include training materials. This just happened upon the realization that the results would be helpful.

There was an initial lack of understanding when this study was presented to the IRB as it was submitted as if it were a clinical trial. The differences between clinical trials, the risks associated with conducting the study, and study documentation were discussed and documented thoroughly.

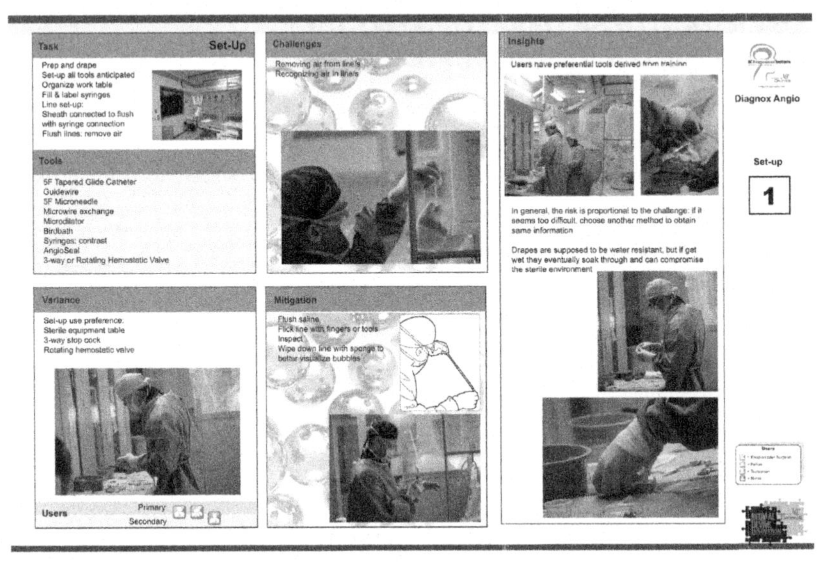

Figure 8.17 Step one: Set-up devices and room.

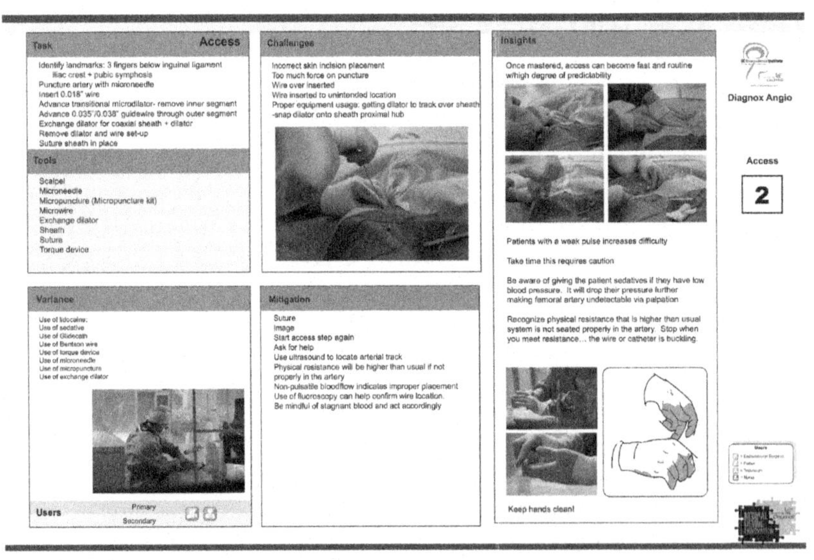

Figure 8.18 Step 2 Gain groin access.

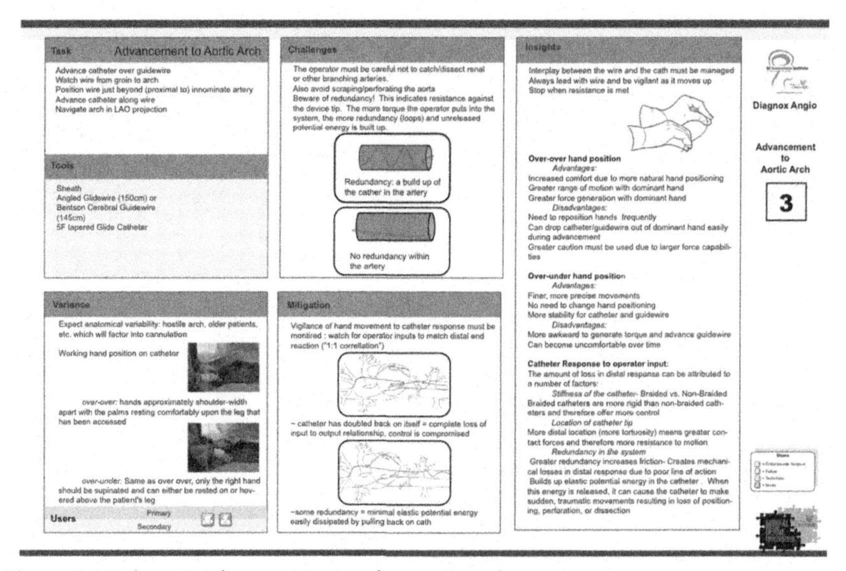

Figure 8.19 Step 3 Advancement to the aortic arch.

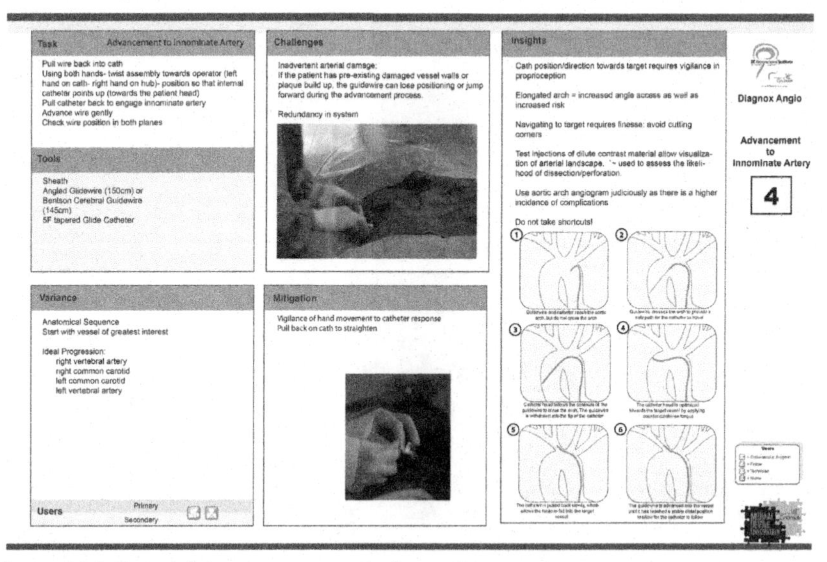

Figure 8.20 Step 4 Advancement to the innominate artery.

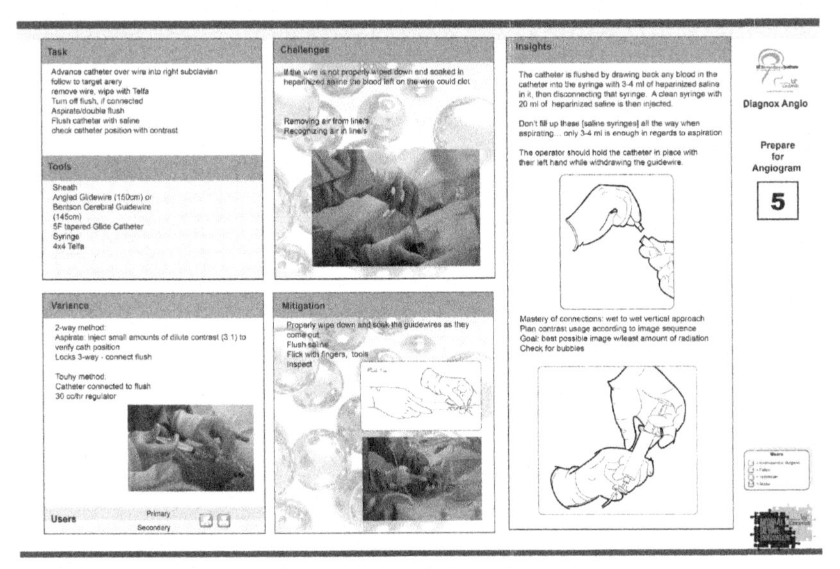

Figure 8.21 Step 5 Prepare for angiogram (imaging).

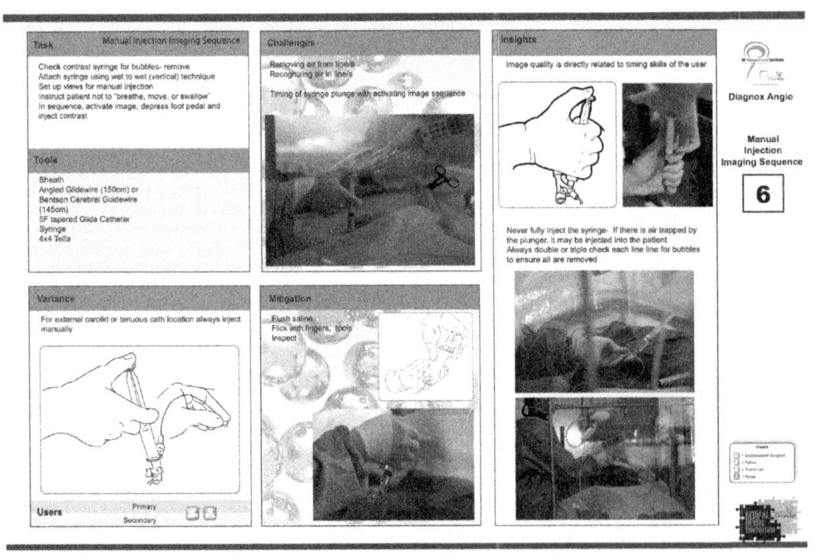

Figure 8.22 Step 6 Manual injection imaging sequence.

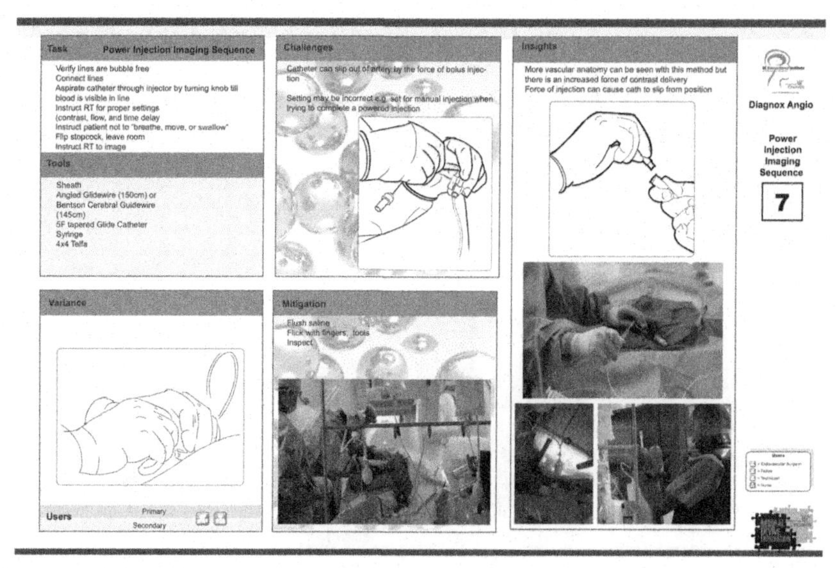

Figure 8.23 Step 7 Power injection imaging sequence.

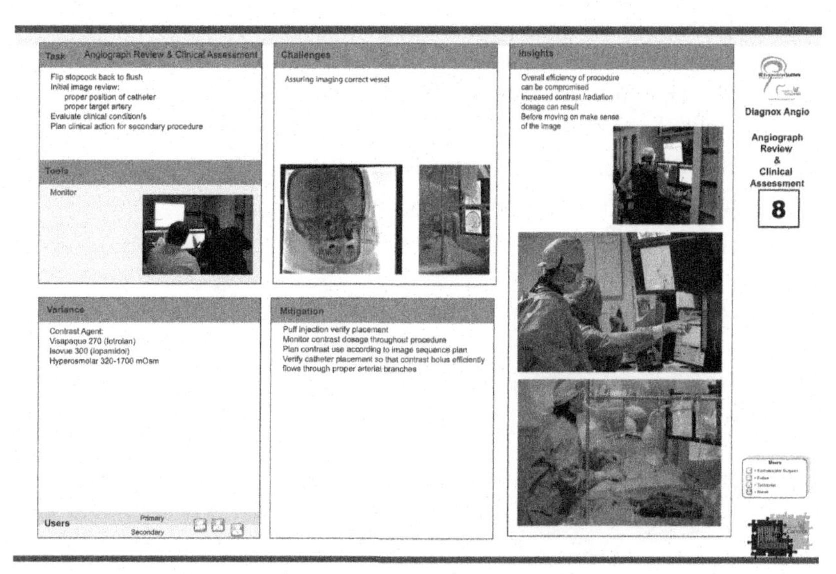

Figure 8.24 Step 8 Angiograph review and clinical assessment.

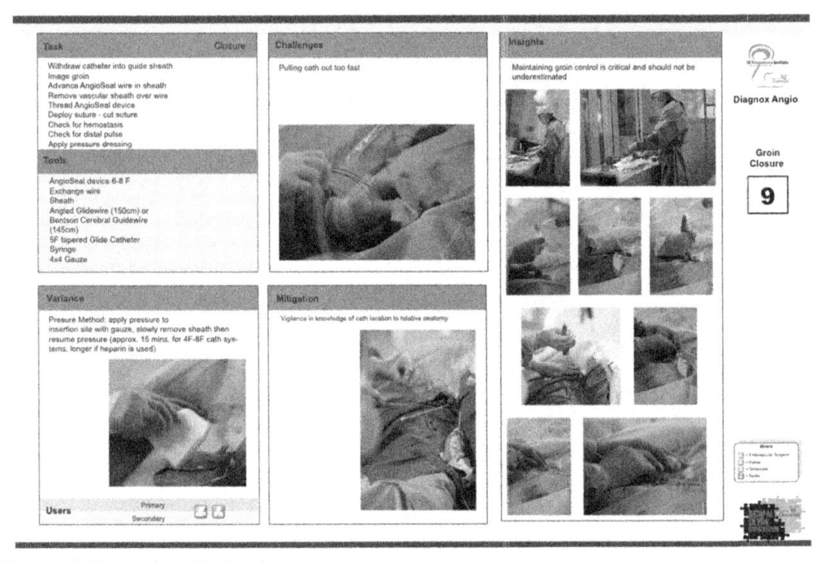

Figure 8.25 Step 9 Groin closure.

8.8 CONCLUSIONS

This study proved a huge success. It involved a large research team over the course of 1 year, then subsequently produced work for five undergraduate students, two graduate masters' thesis projects, and supported medical device innovation teams. Other significant impacts include:

- IP disclosures and two provisional patents filed.
- The relationships gained from this experience carry on today.
- Procedure mapping and systematic analysis provides clinicians confidence in the procedure, maximizing efficiency, and minimizing the potential for errors in training.
- Training materials for national endovascular training course.

Overall this program represents clear collaboration between industry partners and an academic program in order to advance practice while educating future medical device professionals and clinicians.

CHAPTER 9

BioMarker Strategies SnapPath® Case Study: Design Research Program to Integrate a Breakthrough Diagnostic System into the Clinical Environment

Tor Alden
HS Design, Inc., Gladstone, New Jersey, USA

Contents

9.1 INTRODUCTION

BioMarker Strategies (BMS) initiated a program with HS Design, Inc. (HSD) to assist in development of the SnapPath® system (Figure 9.1). The research phase of this program was designed to both familiarize the design team with current technological knowledge and to also establish a comprehensive understanding of the user group and their potential interaction with the SnapPath® system. The research served as the foundation

M.B. Privitera: Contextual Inquiry for Medical Device Design.
DOI: http://dx.doi.org/10.1016/B978-0-12-801852-1.00009-5

Figure 9.1 SnapPath® 1000 and consumable product.

for the development of the commercial system including an instrument and disposable sample cartridge.

A common problem in the new product development process is to identify potential conflicts moving from a technical research environment to that of clinical and other commercial environments. Breakthrough systems like SnapPath® tend to come out of R&D and scientific community. Typically, early stage development has been focused on clinical results based on instrument and assay formulation. Early stage development rarely considers all commercial and user needs, but focuses on results leading to validating scientific claims. Workflow and processes are developed in a research-only environment insulated from real-world conditions for scientific data retrieval. Transition to a commercial product requires the understanding of the users or "actors" involved in sample preparation, processing, and maintaining the system. Moving from subject matter experts in a research setting to serving multiple lab technician users in a clinical laboratory setting requires an understanding of users, uses, and use environments.

This case addresses a process to uncover the development needs of a first of its kind automated live-cell processing system designed to enable highly predictive tests for targeted drug therapies. BMS developed the breakthrough technology to allow biomarkers to be interrogated in live cells outside the human body (*ex vivo*) using fine needle aspiration (FNA) or other types of biopsies. Traditional pathology relies on analyzing dead tissue cells by placing the tissues in formaldehyde or other fixatives for long-term preservation before analysis.

The SnapPath® system required an appropriate design for multiple settings and markets including preclinical academic research, companion diagnostics, and (Bio) pharmaceutical drug development.

The interrogation of live tumor cells provides direct information regarding the responses of signal transduction pathways (pharmacodynamics changes) to targeted therapies, allowing an accelerated path for the identification and development of new drugs. The system can assist biopharmaceutical/pharmaceutical companies in drug development by allowing multiple drugs to be tested on a single sample, prior to initiation of expensive clinical trials to select which lead candidate may have better response rates and assist in enrichment strategies to increase proportions of responsive patients in clinical trials.

While the (bio) pharmaceutical market remains a strong pathway for SnapPath® system, this case focuses on the adoption, usability, and integration of SnapPath® in hospitals for the clinical market where the SnapPath® system would be used in treatment decision-making.

Key takeaways from this case will be the process used to interview key users and to understand the environment and workflow that will lead to successful acceptance. The importance of the user research phase includes:

- Establish the context of use by combining existing knowledge base in user interface, product design, and human factors with the knowledge gained from the user group through site visits.
- To complement the usability engineering process. IEC 60601 third edition recommends completion of the design history file that includes use and usability ("Usability Engineering Process"). The compliance of use and misuse is checked by inspection of the results of the "Usability Engineering Process." Use errors caused by inadequate usability have become an increasing cause for concern.
- To assist in the development of user requirement specifications, product requirement documents, and software requirement documents (URS, PRD, SRD).
- Facilitate a common vision for the SnapPath® system among team members that will serve as a foundation throughout subsequent development work.
- Create buy in with key stakeholders and to best fit into existing protocols.
- Develop useful documentation for making development trade-offs.

Key participants in research included:

Radiologist: Radiologist is only responsible for acquiring the specimen
Sonographer: Locates tumors via ultrasound
Pathologist: Determines cell viability, preserves, and analyzes
Lab Technician: Typical users of lab instruments
Nurses: Front line in patient handling

9.1.1 About the SnapPath® System

SnapPath® is a tissue-based molecular diagnostic testing system which enables functional profiling of live tumor cells from biopsies to guide cancer drug development and treatment selection. The system includes:

1. Instrument
2. Cartridge
   - Onboard reagents
   - Onboard disposables
   - Sample receptacle
   - Microtube outputs

9.2 PROCESS

HSD has developed a research methodology specifically developed for transitioning highly technical scientific programs to commercialization. The foundation is the realization that there are two parts to understanding the successful make up of a complex system: supply and demand. The supply side references the developer's technical challenges, constraints to functionality and marketing requirements. The demand side references the user's needs and use environment. The research process keeps a focus on identifying the value proposition of the customers as well as the technical requirements of the system. In the end, a summary leading to a final understanding of trade-offs and modifications required is developed. A sample of this is illustrated in Figure 9.2.

9.2.1 Focused Innovation

In order to fully understand the value of a research program in complex diagnostic devices, a quick definition of the development process is required. In most cases laboratory instruments are developed in a certified ISO 13485, quality management environment. Based on this, a phased development methodology to satisfy design control is required.

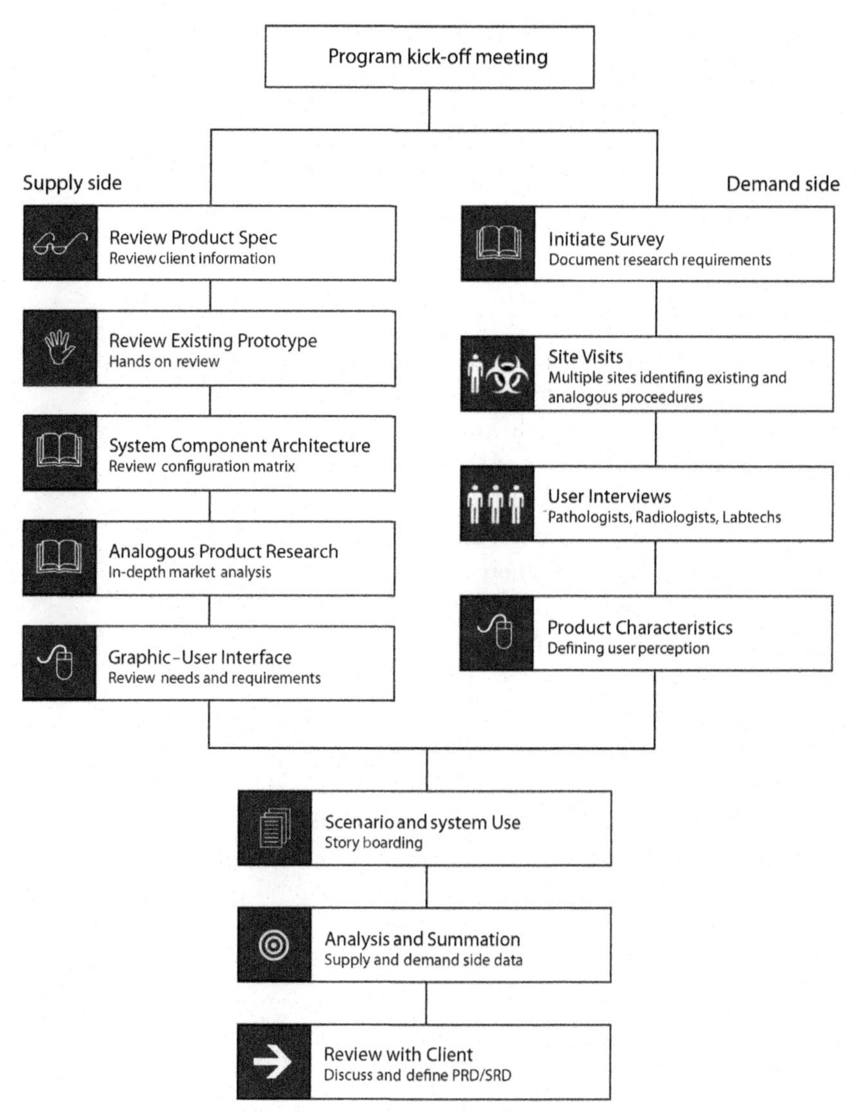

Figure 9.2 Research overview.

Additionally most in vitro diagnostic (IVD) and laboratory equipment falls into a FDA class II category. The primary standard governing medical device design is IEC 60601-1 (Medical Electrical Equipment—Part 1: General requirements for basic safety and essential performance). Often referred to simply as "60601," compliance with the standard has become a *de facto* requirement to bring new medical devices to market in many countries. In particular, the recent shift to the third edition focuses on much greater interaction between the manufacturer to provide verification of use and usability. Risk management, now a vital part of the standard, is a multifaceted, multistep process. Contextual Inquiry (CI) or Usability research provides critical usability validation via formative and summative testing.

9.2.2 Understanding the Phased Process

Understanding the development process is important when performing the user research. Depending on the level of maturity of the technology or company there will be a difference of how much knowledge of the users, process, or environment is available. Established companies with next generation products typically have marketing data such as voice of customer (VOC) and an understanding of use protocol. In new start-ups and new breakthrough products, such as in BMS, there is no predicate device or user protocol to leverage.

HSD has developed a unique approach to work with the scientific community utilizing a "focused innovation" process. There are three pillars that form the foundation of this process: a focus on user-centric design, open collaboration with all stakeholders, and minimizing development risks throughout the project. Adhering to the standardized development methodologies under the guidance of ISO 62366 and AAMI HE75 we examine the product's use under full design control.

9.2.3 HSD's Phased Gate Approach

Phase 0—Strategy and Planning: Typically a company will initiate a phase 0 prior to initiating commercial product development. This will include traditional VOC, minimal viable (proof of product) product, regulatory and project planning, etc. Additionally, intellectual property understanding and competitive assessments will be completed.
Phase 1—Research and Definition: This is where the product moves into early product realization requiring design inputs, user research, system architecture, and proof of concepts (POCs). It is during this phase where CI, identification of stakeholders, user requirements, and preliminary usability testing occurs. These steps include CI, observational, heuristic, tech assessments, and failure mode and effects analysis (FMEA).

Additionally, concept ideation for formative studies such as wireframing and storyboarding, study models, and POCs also occur in this phase.

Phase 2—Development: During this phase, the traditional industrial design, UX, human factors and engineering is initiated, leading to design verification through Alpha level prototypes. This is typically where design control is initiated. Simulated software and prototypes will be developed for formative and regulatory testing.

Phase 3—Preproduction: Detailed documentation, assembly instructions, beginning beta builds, and pilot production occur during this phase. Typically, this is initiated with a manufacturing review and design freeze. Additionally, summative testing and usability protocols along with other agency testing, such as UL, CE, FCC, occurs in this phase.

Phase 4—Realization: Design transfer and release for manufacturing are initiated during this phase, leading to production ramp and finally full market launch.

9.3 USER AND ENVIRONMENT IMMERSION

The team began the research by creating a plan to observe the daily activity within radiology and pathology suites to understand the workflow, individual personnel involvement, patient processing, and existing protocols. We investigated the current process for procuring and processing FNA/Core biopsies in multiple hospitals and outpatient clinics. For this particular use, the initial assumption and client preference was to locate the SnapPath® instrument as close to the sample extraction site as possible. This would presumably mean the device would live in or near the Radiology suite.

Key to the observation was to watch the users interact with each other and establish patterns in the various sites. The research team consisted of 2–3 people at each site visit. Roles were determined prior to visits and research protocol established. While one would initiate questions, the others would take notes, photograph if allowed, sketch relevant features, and generally note observations via ethnographic principals.

9.3.1 User Group

Demand side research further established the user group of the SnapPath® system distinguished by the operations they perform.

- *Oncologist* requests the test and utilizes the results to better treat their patient. The oncologist will be the primary driver of demand.

- *Radiologist* procures the sample and may deposit it into the sample receptacle.
- *Pathologist* may procure the sample if the lesion is palpable, but would oversee the use of the system in their lab. The pathologist is a secondary driver of demand. A pathologist also performs tissue adequacy reviews and may deposit the sample into the sample receptacle.
- *Cytotechnologist* is likely to be responsible for the sample in terms of depositing into the sample receptacle and transporting it to the lab for processing. A Cytotech may also be trained to operate the instrument. They work directly in the lab and would be responsible for running clinical samples.
- *Pathologist's Assistant/Super Tech* may also be trained and responsible for operating the system. They work directly in the lab and would be responsible for running clinical samples.
- *Scientist and research community*—preclinical and early clinical studies conducted in academic or in biopharmaceutical and pharmaceutical laboratory environments. These groups have less stringent requirements and were not included in this part of the study.

Figure 9.3 illustrates how a researcher captures key workflow and environmental constraints in the initial sample prep and the users involved.

Providing HIPAA compliance requirements have been met and permission has been granted, photographs may be allowed. These photos can be extremely useful later in the product development process to facilitate a common vision among the cross-functional team. Figure 9.4 illustrates documented notes of a site visit. Note key points are highlighted for quick reference.

Research was carried out to identify critical aspects that may affect the success of the product and strategize ways to eliminate failure. Figure 9.5 illustrates the post site visit white boarding exercise to determine system flow and transfer gates of the sample.

Later, the white boarding exercises are refined to detailed workflow as shown in Figure 9.6. As mentioned earlier, a key assumption was the Point-of-Care Device would be owned and maintained by Radiology, the department responsible for bedside sample acquisition. However, initial user research provided a counter indication that Pathology Ownership of Instrument was preferred. Primarily, the core biopsy or FNA procedure is typically scheduled at defined times during the day and week to allow for Radiology and Pathology to work together. This procedure typically takes

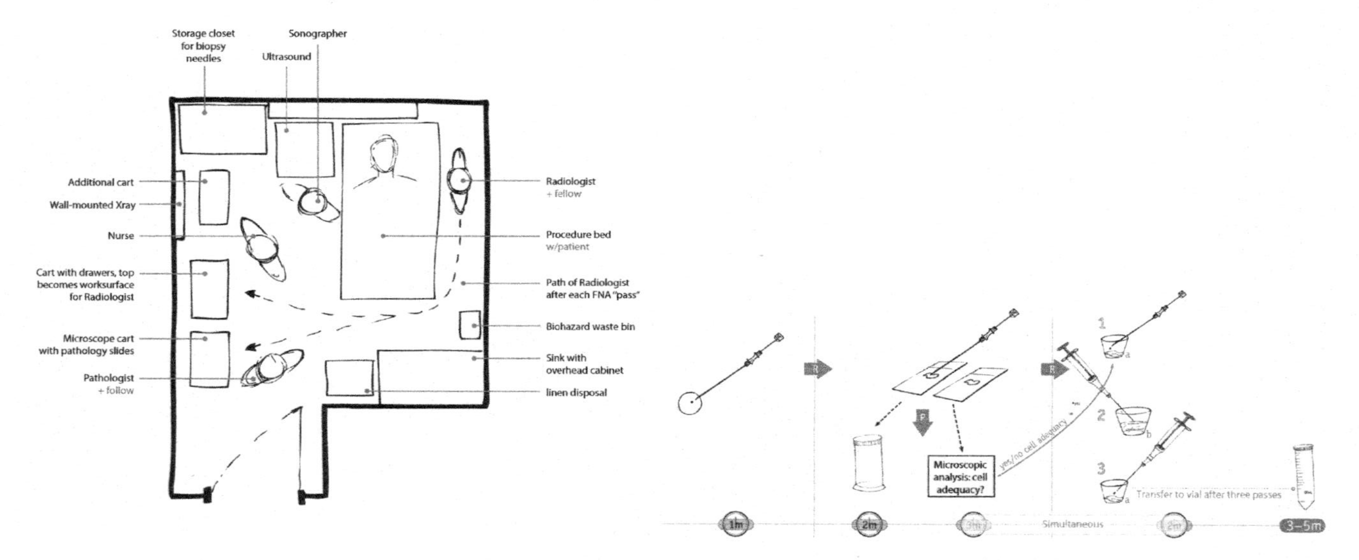

Figure 9.3 Typical Radiology sample retrieval procedure—sketch to detail.

Demand side research—Notes from visit

Site 4

Hospital is an education research institute. This site visit exposed the team to invaluable resources with knowledge in establishing protocols, clinical trials, and molecular diagnostics tests.

Radiology will biopsy samples in various locations including: Interventional, Thyroid Clinic, Breast Imaging, General Ultrasound, and a Cytopathology FNA Clinic. Hospital operates a main cytopathology lab but has two additional auxiliary labs in Breast Imaging and Interventional Radiology that process the volume from those suites.

Key point: Protocol is key for control. It is good to know how much tissue you need up front in a quality, quantity issue. FNAs can be performed blindly (without an adequacy review), immediate adequacy review (at procurement site), or adequacy review after two to three passes. The Protocol should be based on the number of passes, the duration of pass, and the size of the needle rather than an amount of cells because it is hard for people to relate to that value. Protocol should be based on BMS collected data.

It is imperative for the system to be simple and standardized. Based on doctor experience with Genomics, explains to:

- "Dumb it down"
- Quality check the sample up front
- Utilize room-temperature stabilization solution

Key point: One point that was iterated is that clinical validation is critical to the success of the Instrument and to fill a diagnostic niche. The procurement procedure for SnapPath® will be a secondary biopsy so that the sample can be dedicated to the platform. The business model will establish a CLIA Lab and establish multiple trial sights. The location within a trial sight should be an auxiliary lab or Cytopathology.

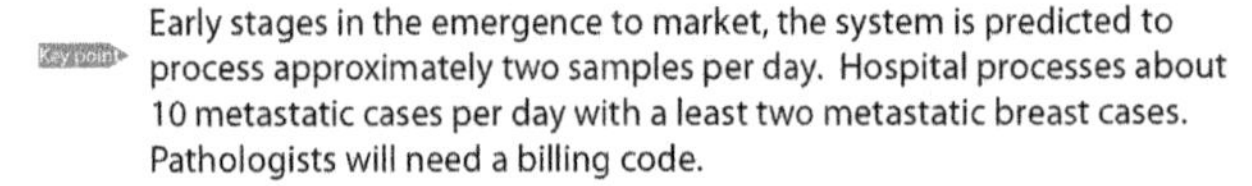

Key point: Early stages in the emergence to market, the system is predicted to process approximately two samples per day. Hospital processes about 10 metastatic cases per day with a least two metastatic breast cases. Pathologists will need a billing code.

Figure 9.4 Notes from single site visit.

approximately 15 min per procedure versus a typical 30 min to one-hour sample processing on SnapPath®.

9.4 VERIFYING ASSUMPTIONS

By concentrating on the user experience and the needs of the environment, we were able to gain actionable feedback in regard to SnapPath® design opportunities. We realized that the initial assumption of locating the device close to the sample extraction site was problematic. Our investigation uncovered product placement opportunities and helped our client understand that their initial assumptions did not work with the presumed workflow.

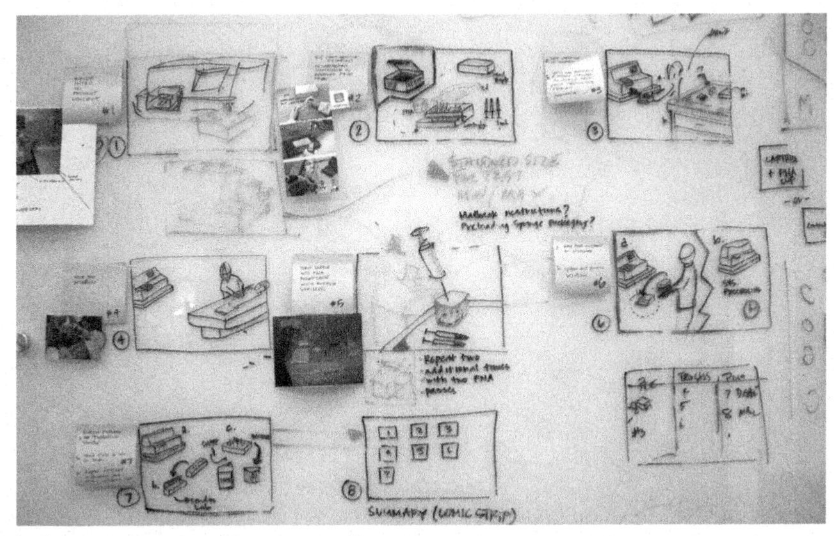

Figure 9.5 White boarding system flow.

Constructing use scenarios is a very effective way of obtaining a realistic picture of the envisioned process. This helps us to gain a preliminary and very important understanding of how the product will perform and how the user will interact with it. Additionally, use scenarios are the first step toward implementing research findings into the design execution phase.

The following section has several maps illustrating the use-case scenarios of the SnapPath® system and the environment in which it could operate. This section will cover (1) the current process observed at the site visits and (2) the proposed use scenarios of the SnapPath® system centered on the different types of input. SnapPath® has three different workflows scenarios depending on type of sample input (see Figure 9.7). Most of this case study focuses on the FNA sample process.

9.4.1 FNA Sample Processing

Understanding how biopsy samples travel throughout the current process enabled us to design a disposable cartridge that best fits that workflow. The Research team developed process maps and the Design team collaborated in developing concept props for user evaluation. It became evident using the existing protocol for sample procurement was ideal. The SnapPath® system, patient tracking of sample, and ownership of sample from Radiology to Pathology transfer were defined. This process was broken into five phases: (0) sample procurement, (1) system sample

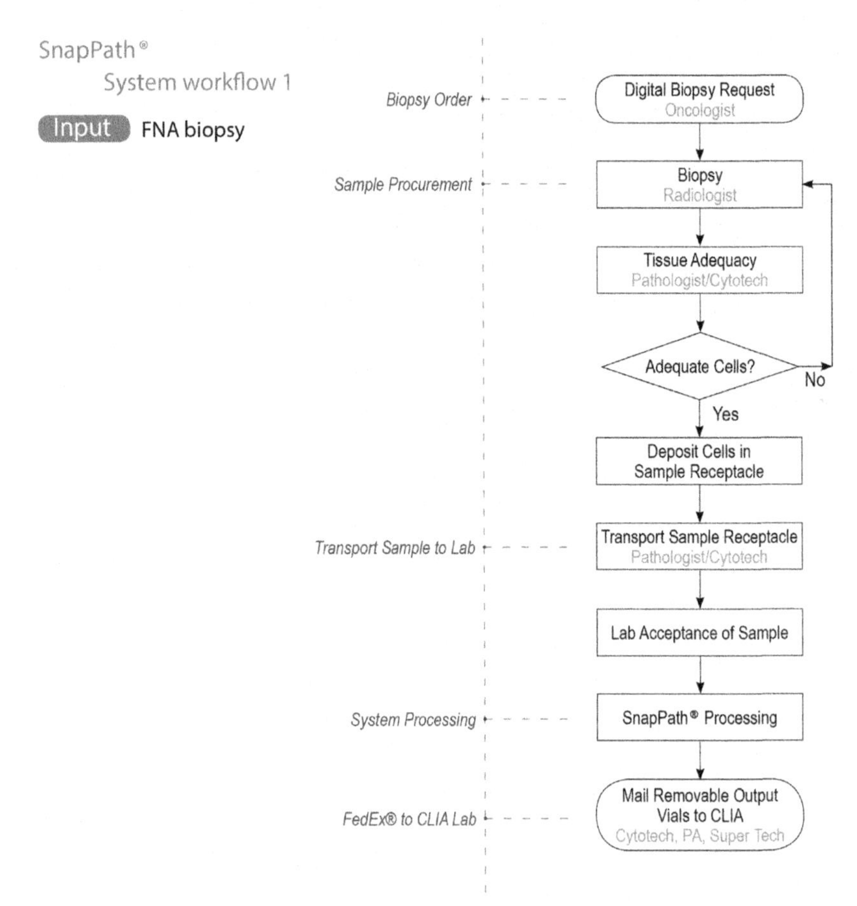

Figure 9.6 SnapPath® workflow processing FNA sample.

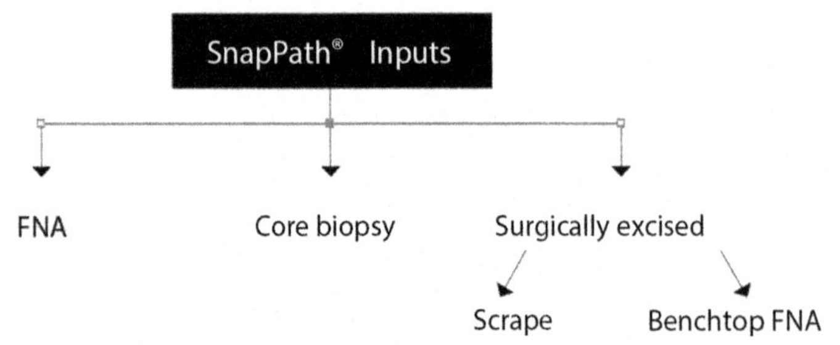

Figure 9.7 Potential biopsy input sources.

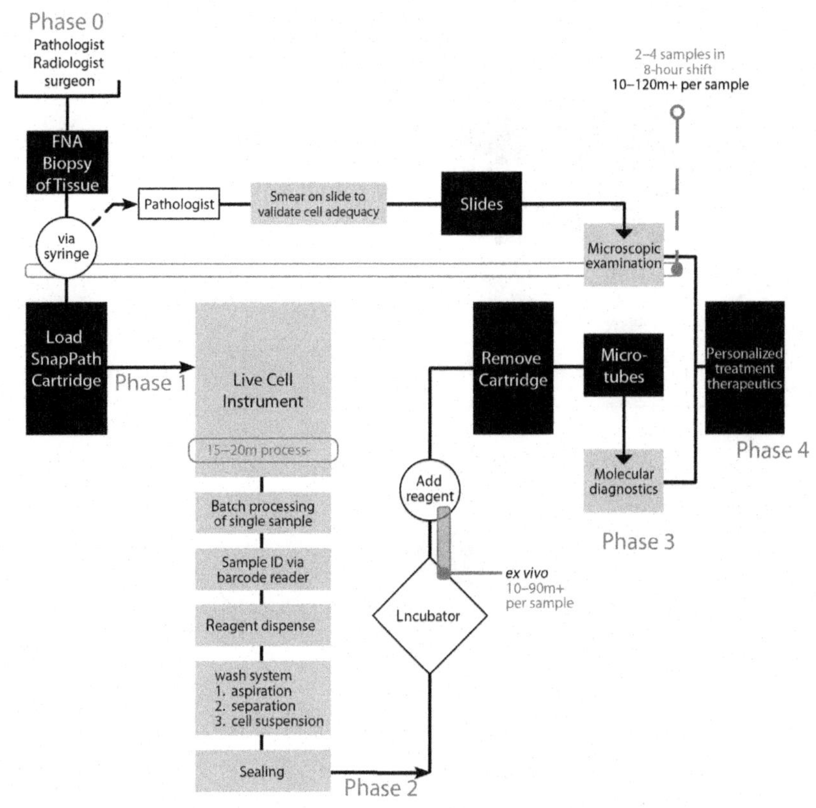

Figure 9.8 Diagram illustrating the SnapPath® system process.

preparation, (2) system sample incubation, (3) molecular analysis at CLIA lab, and (4) oncological personalized treatment therapeutics (see Figure 9.8).

Based on the above workflow, flexibility in the cartridge design to adapt to sample transported from Radiology to Pathology was imperative to the design constraints. The cartridge had become the main interface for the lab technicians. The design of the cartridge was innovative in itself as it contains all the test components (assays, pipettes, incubation wells, sample) eliminating the need for onboard fluidics, disposables, and waste. See Figure 9.9.

9.4.2 The Graphic–User Interface

The UX design was a key component of the system and acts as the user's method for operation. The research team began understanding the

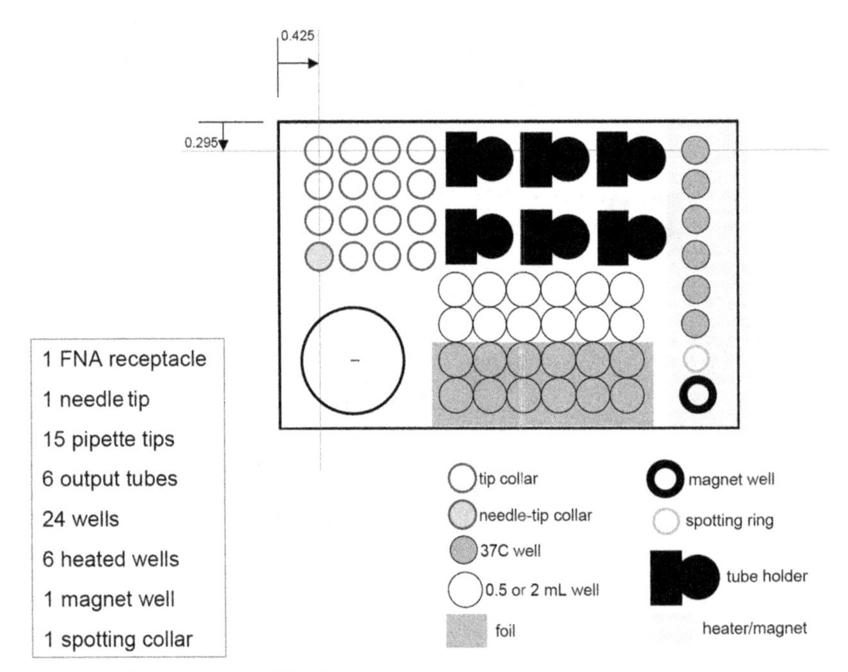

Figure 9.9 Prototype cartridge layout.

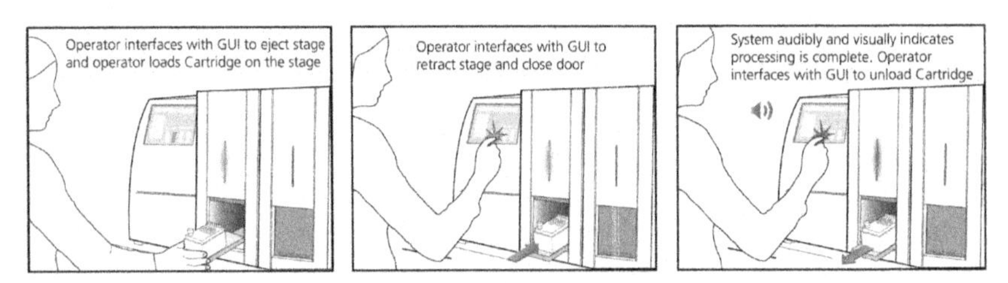

Figure 9.10 Theory of operation.

problem by developing the theory of operation, workflow, wireframe concepts, as well as methods to validate the concept. Figure 9.10 illustrates a storyboard that was shown to laboratory technicians to verify the team's understanding.

With the consideration of user feedback, the design was further developed to enhance the brand identity and perception to create the utmost experience to manage multiple processing cartridges. Based on earlier research that illustrated the need for multiple instruments to handle sample throughput, the design included flexibility to add more modules. Figure 9.11 shows a preliminary UX interface that allows four instruments to work off in one central module.

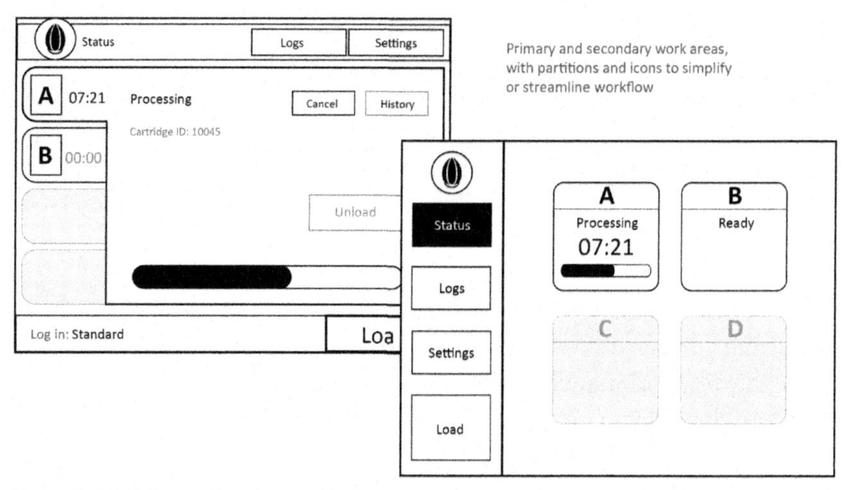

Figure 9.11 UX interface concepts.

9.5 SUMMARY

In summary, the user research allowed our team to understand the process flow of sample retrieval and processing. This knowledge allowed us to uncover major elements that made the SnapPath® more effective in the environment and allowed users to integrate their tasks more seamlessly. The research led to a final understanding of trade-offs and modifications and allowed the SnapPath® system to seamlessly prepare a sample for predictive treatment analysis for targeted drug therapies on an individual patient basis. In order to optimize throughput and cater to the needs of an individual laboratory, the system has the flexibility to expand in capacity. The modular design easily connects additional "modules" for increased efficiency.

The research program identified and integrated key user groups that would be part of the systems success, including:

- Key users in traditional cell retrieval are:
  - Radiologist: Retrieves cells via biopsy or FNA
  - Sonographer: Locates tumors
  - Pathologist: Determines cell viability
  - Technician: Provides sample prep and diagnostic run
  - Oncologist: Determines what tests are required

Additionally, based on these different users and markets, the system needed flexibility to work in the following environments:

- Small research only (product is used minimal times per year or small sample batches)
- Specialty labs: integrated in both small and large batch process
- Clinical and POC
- Central Labs
- Large throughput laboratories
- Academia
- (Bio)Pharmaceutical Companies

BMS's scientific and marketing team, alongside HSD's research team, successfully developed a CI research program to close the loop of how the new breakthrough SnapPath® system would integrate into existing hospital protocols. Further, how the system would best be optimized for the users, starting from initial patient sample retrieval to delivery of CLIA lab results to oncologists. The Supply/Demand research approach is ideal for research or complex system product development as it addresses both the technical and user requirements. The supply side references the developer's technical challenges, constraints to functionality, and marketing requirements.

The demand side research references the user's needs and use environment. By doing a deep dive into BMS's existing prototypes, we were able to provide a system architecture that allowed for expandability and small footprint. The understanding of the technical challenges in sample procurement and reagents allowed the team to develop a flexible system with all consumables on the disposable cartridge.

The supply side research, based on over six leading hospital site visits, allowed the team to identify the value proposition of the users and the environmental issues. Protocols in various hospitals are different. By getting a good cross section, BMS was able to identify the key points in streamlining the process. The research helped identify the specific challenges facing the system interaction, sample acquisition and the hand-off of sample to lab. It further provided needed insight in the interface of the instrument and cartridge.

The combination of understanding the technical and manufacturing needs (supply) and the marketing and user needs (demand) allows for quicker development and time to market. The research was critical in the development of the user product requirements and the product and software requirement specifications. User verification of concepts and system architecture iterations allowed the team to gain buy in and maximize user's efficiency in the instrument workflow. Additionally, the research document was used in the final design history file as part of the FDA guidance documents.

ACKNOWLEDGMENTS

Special thanks and acknowledgments for all the contributions HSD had with the BioMarker Strategies team in putting this case study together, specifically Greg Bertenshaw, VP of R&D, Jerry Parrott, President & CEO, and Dr. Douglas Clark, Co-Founder. Additional thanks go out to our contract development partner, Sparton Medical. The BMS and Sparton team included doctors, scientists, and marketing, engineering, manufacturing, and regulatory groups that worked closely with the HSD team in a true collaborative environment to transform the CI and design research into concise product and software requirement documents, ultimately leading to the development of the Silver MDEA award-winning SnapPath® system.

CHAPTER 10

Contextual Inquiry as a Tool for Medical-Device Development: The Case of Harmonic Focus

Stephen B. Wilcox
Design Science, Inc., Philadelphia, Pennsylvania, USA

Contents

10.1 INTRODUCTION

This is a story about using contextual inquiry to support the development of a new medical device. It is more than that, though. It is also a story about using contextual inquiry to change the product-development process at a major corporation and to change the role of industrial design within that process. Before I begin the story, though, let me say a few words about why we do contextual inquiry in the first place, although I know this risks being somewhat redundant to other contributions to this volume.

From my perspective, the purpose of contextual inquiry is to create a foundation of accurate information to improve the likelihood that a device will be a good fit—a good fit with the people who will use it, with the environments in which it will be used, and with the procedures to which it will contribute. Putting it this way, of course, begs the

M.B. Privitera: Contextual Inquiry for Medical Device Design.
DOI: http://dx.doi.org/10.1016/B978-0-12-801852-1.00010-1

question of why we cannot accomplish this purpose by simply asking the right questions of the right people. In fact, as I will describe below, our version of contextual inquiry is complicated and time-consuming. It involves extensive real-time observation of real events (in this case, surgical procedures), video-recording with a complex multicamera system, and painstaking analysis of the data, thus obtained. Why go to all this trouble when we could simply ask a sample of medical professionals the questions we want to answer—what they want, what they do, and what the environments are that they work in?

The short answer is twofold—(1) that people's descriptions are always incomplete and often inaccurate and (2) that, even if they were complete and accurate, there is no substitute for seeing phenomena directly instead of developing an indirect understanding of them through the accounts of others.

There is, in fact, enormous evidence for the first point—that peoples' descriptions are fallible. For example, the psychologist, Elizabeth Loftus (1996), in her classic book, *Eyewitness Testimony*, described study after study demonstrating that peoples' recollections of directly observed events tend to be highly inaccurate, something that the rememberers are often unaware of. People tend to integrate information from other events, to be effected by discussions about events that follow them, etc. And, if we think about the case of a medical procedure, the medical professionals do not even notice, or have any particular reason to notice, much of what we are interested in—e.g., specific settings, frequencies of events, complex interconnections between chosen procedural options, and so on. And that is not to mention the fact that, for many reasons, people do not always tell the truth, even when they do know the facts.

At any rate, the sort of deep understanding that benefits a device-design team cannot be achieved by discussions alone. There is no substitute for seeing the awkward postures, the near misses, the waxing and waning of stress, etc. Only direct observation can provide this deep understanding.

So, these are the reasons that we initiated contextual inquiry for the project I am going to describe and the reasons that we conduct contextual inquiry, in general.

The particular project I want to describe here was conducted to support the development of an "ultrasonic device," one that eventually was named *Harmonic Focus*. As most readers probably know, an ultrasonic device is a type of instrument that uses high-frequency vibration instead of an electrical current to transect tissue hemostatically, that is, to simultaneously cut and coagulate. The key advantages of harmonics relative to electrosurgical devices are that they are less traumatic to surrounding tissue and generate less smoke, while still transecting tissue with little or no bleeding.

Some years before the project began, Ethicon Endo-Surgery (ETHICON), a J&J operating company, had acquired a company called *UltraCision*, that made harmonic devices. The product line ETHICON obtained consisted of a reusable transducer (the engine) to which were fitted disposable cutting/coagulating instruments. The transducer is the mechanism for producing the high-frequency vibration that is transmitted to a given instrument. After acquiring UltraCision, ETHICON had grown the product line, but their market was largely restricted to applications in general surgery and gynecological surgery. One limiting factor was the size and weight of the transducer, making it difficult to develop instruments appropriate for more "delicate" surgical applications, such as ear, nose, and throat (ENT) surgery and peripheral vascular surgery. Figure 10.1 shows an example of an ETHICON harmonic instrument that existed at the time the project was initiated.

The catalyst for the project described here was the technical feasibility of developing a new transducer—smaller and lighter than the original—designed to allow ETHICON to expand the device's applications (see Figure 10.2).

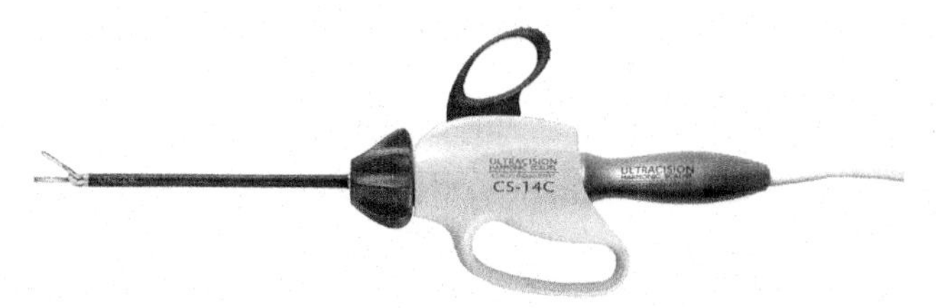

Figure 10.1 A harmonic instrument that existed at the time the project was initiated.

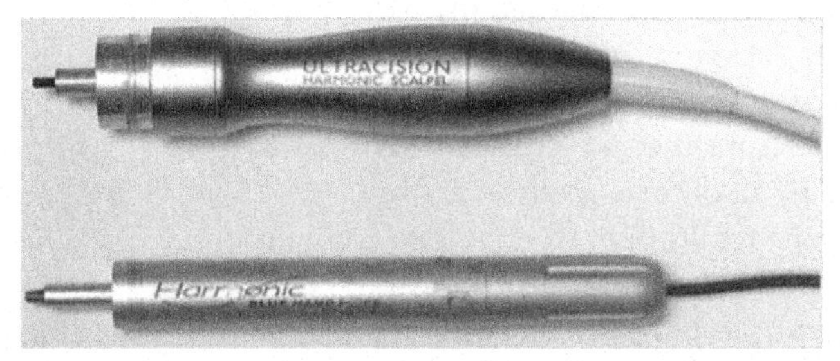

Figure 10.2 The original transducer (top) and the newly developed transducer (bottom).

The immediate problem was to determine which new applications to address first. ETHICON wanted an application that was a likely candidate for a harmonic device (e.g., requiring multiple vessel ligations) that was performed frequently enough that some reasonable volume of devices could be expected, and that would lend itself to use by ENT surgeons. Using these criteria, ETHICON chose the *thyroidectomy* procedure to focus on. Their approach of choosing a specific procedure was in line with a notion mentioned by Alan Cooper (2004): that (to paraphrase), "it's more important to be precise than to be accurate." The notion was that focusing on a very specific procedure would force the team to address all the necessary little details much more thoroughly than would be the case if the application were defined more vaguely or broadly. The thought is that, if an instrument can be made to be really good for the target procedure, it is likely to be good for other procedures as well.

Our mission, then, for the project discussed below was to provide an understanding of thyroidectomy procedures for the team that was charged with developing a new harmonic device for those procedures.

10.2 METHODOLOGY

10.2.1 Planning

We started where we always do: by trying to understand, in as much detail as possible, what the design team wanted to know and how the desired information could be presented to be used most effectively. We find that if we start at the end, so to speak, and work backwards, that we are less likely to go astray. In other words, a clear, shared vision of the end point provides a guide that controls all the little decisions that have to be made throughout a project. This is the point at which we determined three key things:

1. Since ETHICON's design teams are interdisciplinary, the users of our information came from various backgrounds, including engineering, industrial design, clinical practice, and marketing.

   Thus, it was crucial to make sure that we provided information that could be easily understood by all of these groups.
2. Some users were likely to use our results only to drive key decisions; others were likely to "wallow" in the details.

   It followed that it was important to provide a hierarchy of information—straightforward, actionable conclusions, as well as details that allow those who so desire to engage in a "deep dive."

3. Our information was liable to be used in two fundamentally different ways—by groups in meetings and by individuals at their own work stations.

   The key implication of this fact was that we needed to create deliverables that were optimized for each of these scenarios.

The next step was to tackle the planning and logistics of the study. There is not space here to go into the details (each of which could be its own case study), so I will just list some of the things that had to be done at this point, as they do for any such study, and provide a few of the relevant details:

- Determination of the characteristics of the sample—types and locations of procedures, types of surgeons, etc.—and creation of a formal screener that reflected those characteristics.

  Examples of variables that make a difference are the surgeon's experience, the surgeon's gender, the type of hospital (e.g., large teaching hospital versus community hospital), position relative to current segmentation of the market, and the type of devices presently being used.
- Design and execution of a recruiting strategy (hopefully, as in this case, with ETHICON obtaining agreements from surgeons).

  Obtaining permission from surgeons is just the beginning of a process that involves such things as acquiring additional permissions from hospital administrators, verifying vaccinations for the research team, mandatory infection-control training, etc. One difficulty is that every hospital has a different set of requirements.
- Creation of a research protocol that specified, in detail, what we wanted to observe (and video-record) and what we wanted to discuss.

  Without an adequate protocol, observational research is "just watching" rather than true research that can be called *contextual inquiry.*
- Creation of required forms covering issues such as informed-consent, video-release, and nondisclosure.
- Education of the research team.

  In order to be effective, the field researchers had to know, in great detail, what they were going to observe—what the relevant anatomical structures were, what would happen to those structures, what the clinical goals were, what the instruments were that would be used, etc. This knowledge began with a "boot camp" from ETHICON, followed by intensive study of our trusty Gray's Anatomy (Standring, 2008), and review of videos of the procedure, as well as relevant articles from medical journals. This education was crucial since the observers had to be able to understand and report their observations in a clinically relevant way (e.g., "transecting Berry's ligament" rather than "cutting something"). It was also necessary to understand the procedure in order to ask useful questions of the surgeons (e.g., "Why did you start by exposing the left lobe?" versus "What are you doing now?"). It is also important to have a deep understanding of what will unfold in a procedure in order to optimize the video setup.

Once we had finished these various tasks, among others, we were ready to begin conducting the research.

10.2.2 Conduct of the Research

For this project, as for most such projects, we sent a two-person team into the field—a researcher and a videographer. The researcher had training and experience in observing and interviewing. A separate videographer was necessary because, for one thing, we used a three-camera configuration with a separate microphone on the surgeon. In this particular case, we used two tripod-mounted cameras (one to capture a close-up of the surgical site and one that captured a wider view showing the medical professionals involved and the equipment table) and a handheld camera that provided the flexibility to capture various phenomena that emerged in the course of the research. We added a wearable microphone on the surgeon to assure that we could clearly hear what he or she said, which is not always possible from the audio component of video cameras.

Figure 10.3 shows a typical setup, in this case from a more recent project, since our technology has evolved subsequent to the Focus project. The image shows three cameras: an overhead boom (upper left), a handheld camera on a monopole, and a tripod-mounted camera (far right, with, in this case, a researcher monitoring it). It is also common with this type of research to add a direct feed from any imaging that is used.

The job of the researcher was to interview the surgeons before and after the procedures (and during the procedures, if possible, which it usually was) and to carefully observe and take notes.

We observed a total of 23 complete thyroidectomy procedures in the US, Italy, Germany, and France. For each observation, we arrived prior to the start of the procedures to set up our equipment and talk to the

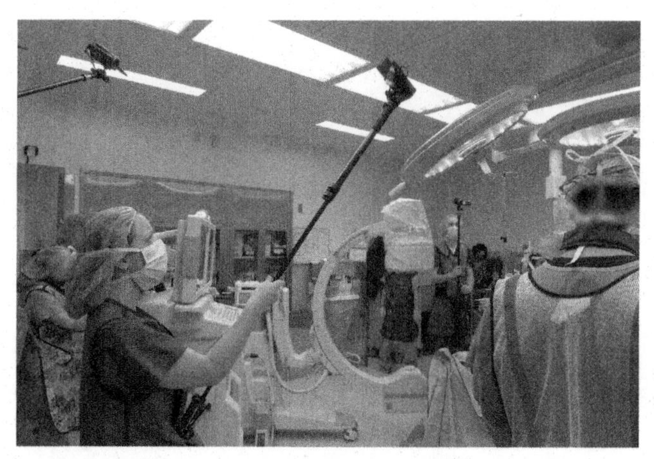

Figure 10.3 A three-camera setup in an operating room.

medical team, and we captured everything we could from the early preparation to clean up of the room.

10.2.3 Analysis

We came away from the field with many hours of video and audio recording and hundreds of pages of notes. Making sense of all this is, perhaps, the biggest challenge of contextual inquiry. A common and relatively easy approach is to draw from the intuitive understanding that the researchers develop over the course of a project and to focus on interesting anecdotes to draw conclusions from them—examples of surgeons struggling with a step of the procedure, for example. Such an approach is certainly helpful and always included as one component of our analysis. However, contextual inquiry can also yield comprehensive data of various types. In this case, we used the video to perform a detailed analysis of timing, instrument use, etc. Our video analysis involved determining, at each short interval, which instruments were being used, who the users were, what was being done, etc., and entering these data into a spreadsheet.

For the researchers' field notes, we translated all of the insights and factual observations into categories of interest to the team and also entered these data into spreadsheets.

One key to an optimal analysis is to assure that every single piece of potentially useful information gets reported in some form that is

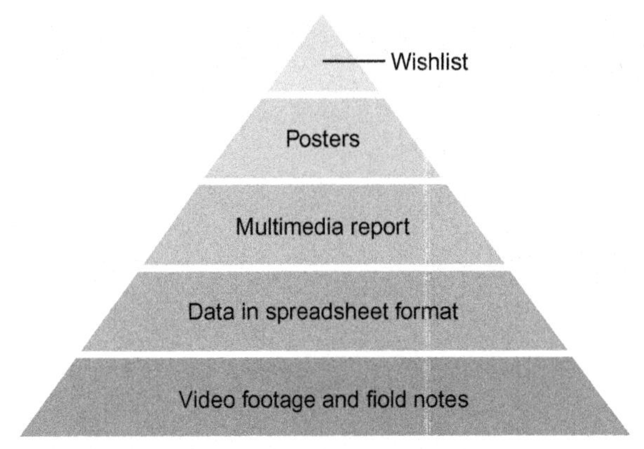

Figure 10.4 Information pyramid for the Focus project.

accessible to the users. Making such complicated information usable is the subject of the next section.

10.2.4 Reporting

I am an advocate of providing information as a "pyramid"—simple and direct answers to key questions at the "top" and "raw" data at the bottom (see Figure 10.4). Let me start, then, at the bottom and work up.

The "least refined" data were the video footage and the notes taken by the researcher. We provided both of these in a "cleaned up" form to provide anyone who so desired the opportunity to "wallow" in them. "Above" that were the spreadsheets with the results of our counts of instrument exchanges, timing and other measurements, frequency counts from interviews, etc. We find that these spreadsheets tend to be used when there is a specific quantitative question that requires an answer (e.g., "What was the longest single period of use of instrument x")—sometimes months or years later, as new questions emerge.

The next level up was the multimedia report. This is where we summarized key insights—e.g., problems to solve, product requirements—illustrated by video clips, figures, graphs, and so on. This document is the closest we typically come to a conventional report. We have found over the years that thick, text-heavy reports do not tend to be used effectively by product-development teams.

Next were posters. The idea of a poster is to present information in a format that can be physically put on the wall. This makes the information instantly available in meetings, where many of the decisions are made,

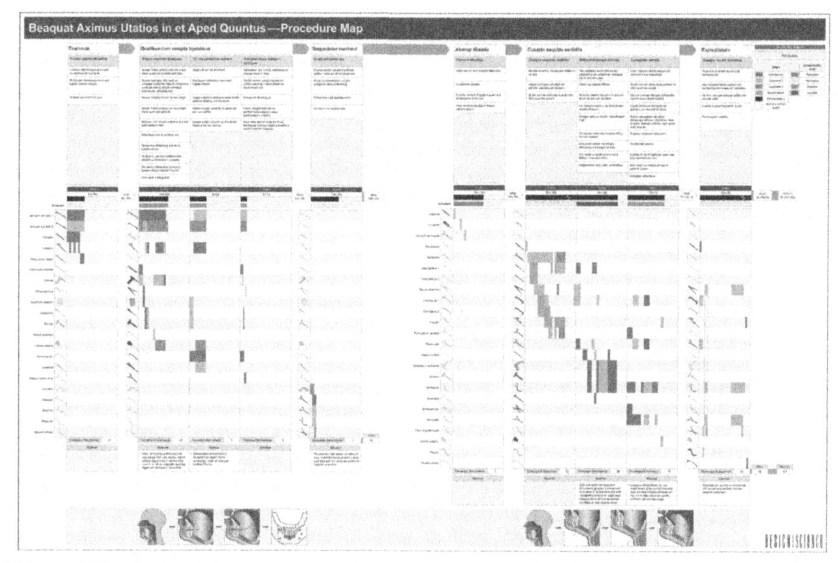

Figure 10.5 A procedure map.

and it avoids the "keyhole effect" associated with electronic images when there is a lot of information to be presented. Figure 10.5 provides an example of a type of poster that we call a "procedure map." We have intentionally obscured the text and reduced the size to protect the information contained in it, and the map and the other examples here are not actually from the Focus project, since the information remains proprietary, but the figures provide the types of posters we provided on this project.

The issue *vis-a-vis* the "keyhole effect" of an electronic format is that projecting the whole image requires that the text be too small to read. Thus, in order to read the text, it is necessary to only observe one portion of the image at a time (through a "keyhole"), which makes it hard to see the "big picture." The way around this problem is to print it out as a large-format poster. The procedure map shown in Figure 10.5 is designed to be printed out 3–4 ft wide.

The procedure map provides a procedural summary that can be understood "at a glance." This one has time going from left to right, and each of the rows in the main body of the figure represents the time that a particular instrument was used. Each column represents a phase of the procedure, and the larger colored bars above the "instrument rows" indicate the estimated stress level of the surgeon for that phase (green for low

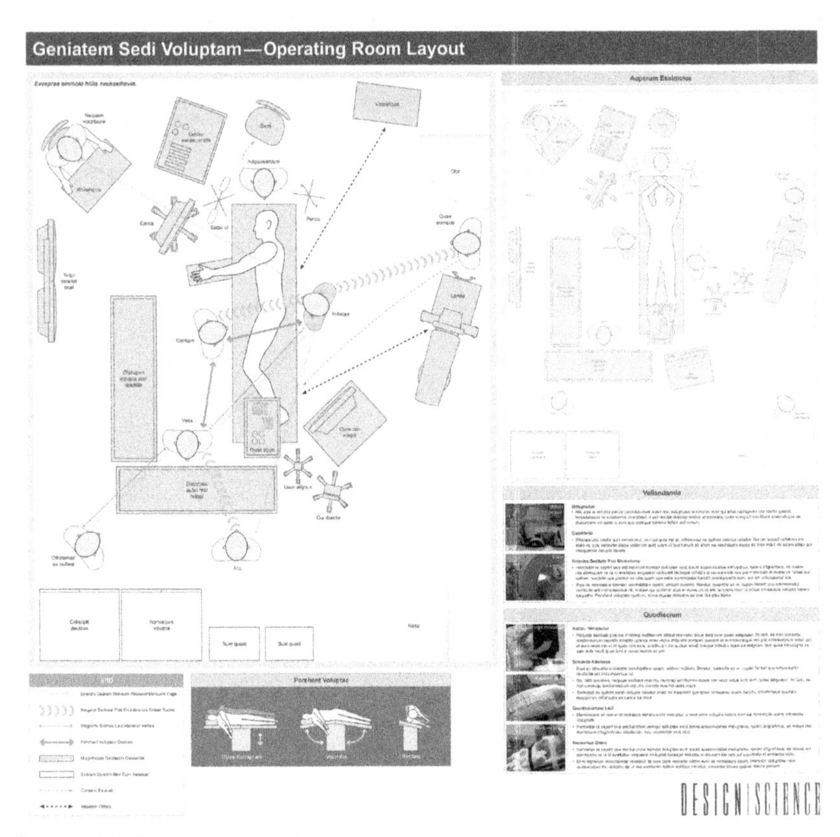

Figure 10.6 Operating room layout.

stress, red for high stress). The text at the top summarizes the substeps; the text at the bottom provides comments regarding particularly noteworthy observations. The images at the very bottom provide a clinical summary in a storyboard format to orient the user to what is actually happening clinically in the procedure.

The procedure map represents the use of "information graphics" (cf., Tufte, 2001). Information graphics are a tool for making complicated information accessible.

Figure 10.6 shows another example of information graphics used to create a poster. It is a plan view of an operating room showing the typical placement of equipment and people, with patterns of communication and patterns of instrument exchanges. The larger layout is the most common pattern. The smaller one in the upper right provides an alternative seen

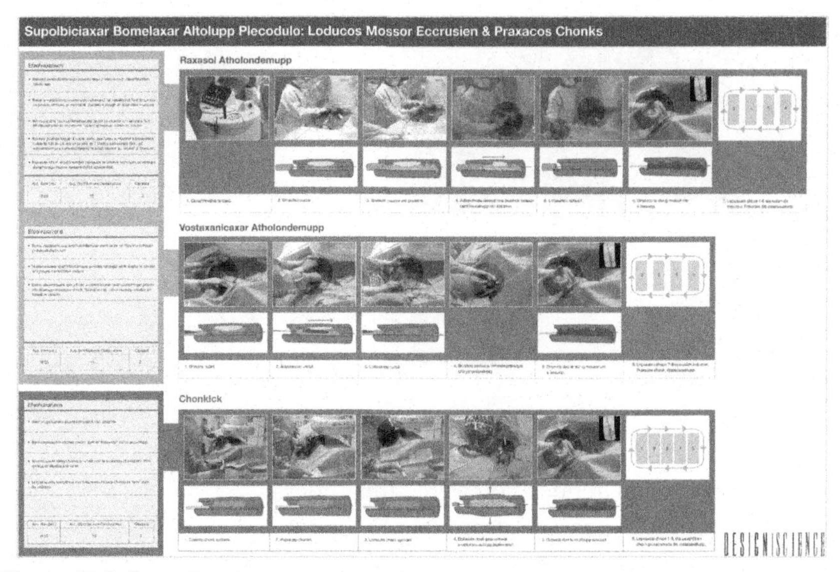

Figure 10.7 Scenario.

less frequently. As with the procedure map, we provided additional information, as relevant.

Figure 10.7 shows one more example of a poster. This is what we call a "scenario," which uses a storyboard format to capture the steps of a subcomponent of a procedure, typically when there are multiple alternatives. They can either be done photographically or via illustrations. This example shows three ways that a particular task was done.

Finally, at the very top of our information pyramid, was a simple, prioritized list of the basic requirements for the new device that we obtained from our research. Of course, it was built on the foundation of the other deliverables that we produced.

10.3 CONCLUSION

The ETHICON team used the results of our contextual inquiry to develop Harmonic Focus, as shown in Figure 10.8. The contextual inquiry was used to inform the overall configuration (which, as can be seen by looking back at Figure 10.1, was dramatically different from the existing product line), many of the ergonomic characteristics of the device, and subtle details of the "business end," the jaws of the device. Only by understanding, in great depth, how ENT surgeons used their

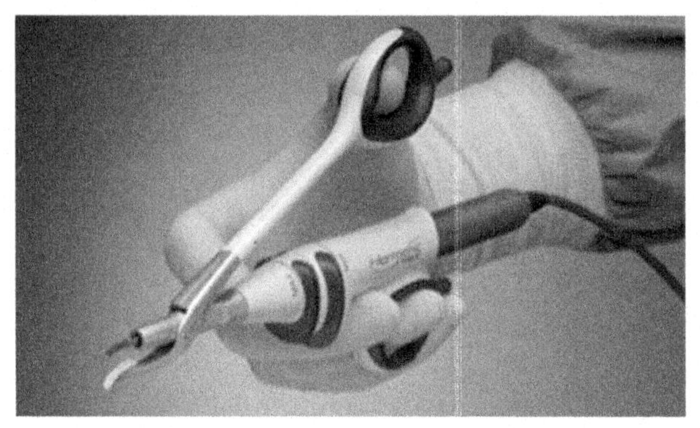

Figure 10.8 Harmonic Focus.

existing instruments, what their biases and tendencies were, the other instruments that Focus would be used with, etc., was it possible to "get it right."

Of course, the contextual inquiry was only an early step in a several-months-long development process that involved a number of rounds of design/prototyping and usability testing (among many other things) to work out all the details and assure that the device was optimized.

Upon its introduction, the product was a great market success, and we were all proud of winning a Gold Industrial Design Excellence Award for the project. As this is written, Harmonic Focus has expanded into a product line with a number of alternative configurations and continues to do well in the marketplace.

The other important result is that the project demonstrated the value of contextual inquiry within ETHICON, so changed their approach to product development. Part of this is the changing role of industrial design. In the past, industrial design had an important role within the company, but there was some tendency by some development teams to first design a working device, then bring in the designers to assure a successful transition from the working "breadboard" to a real product.

An advantage of contextual inquiry is that it is "owned" by the industrial design team, so the designers have become the providers of not only design services, but also of key information that development teams recognize as valuable to their efforts—bringing industrial design in at the onset of projects. ETHICON has gone on to conduct contextual inquiry for many different procedures and many different product lines. Also, the results of contextual inquiry, particularly the posters, have found multiple other uses:

1. As training tools, to help orient surgeons learning to use devices how those devices are used within procedures and to educate new employees.
2. As vehicles to discuss the needs and preferences of surgeons in greater depth than is possible without the procedural details provided by the posters.
3. As a means to assess relevant marketing claims that relate to surgeon instrument use or patient outcomes.
4. As tools for uncovering opportunities for future product development.
5. As tools to help set project/product requirements, targets, ranges, and metrics for team and/or concept success.

I believe it is fair to say that contextual inquiry has become an important strategic tool for ETHICON's product development, one of the reasons that ETHICON remains the world leader in surgical instruments.

So the story has a happy ending.

ACKNOWLEDGMENTS

I would like to thank and acknowledge the contributions of the industrial designers with whom Design Science collaborated on this project. Stuart Morgan, ETHICON's Design Director at the time, provided much of the vision for the whole approach to contextual inquiry represented here. Steve Eichmann and Matt Miller, truly talented industrial designers on the project (and both now promoted to more responsible positions), worked closely with us and contributed in many ways to our efforts, which were very much a true collaboration between Design Science and ETHICON.

REFERENCES

Cooper, A., 2004. The Inmates Are Running the Asylum: why high tech products drive us crazy and how to restore the sanity. Sams Publishing, Indianapolis, IN, USA.

Loftus, E., 1996. Eyewitness Testimony. Harvard University Press, Cambridge, MA, USA.

Standring, S. (Ed.), 2008. Gray's Anatomy. Elsevier, Amsterdam.

Tufte, E., 2001. The Visual Display of Quantitative Information. Graphics Press, Cheshire, CT, USA.

CHAPTER 11

Using CI to Inform Design Development of an Incision and Drain Packing Device for Use in Emergency Medicine*

Mary Beth Privitera
University of Cincinnati and Know Why Design, LLC, Cincinnati, Ohio, USA

Contents

11.1 INTRODUCTION AND BACKGROUND

An abscess is a subcutaneous collection of pus caused by an obstructed sweat or oil glad, an infected hair follicle, or debris. These cause the body's inflammatory response to activate in an attempt to contain the infection. An abscess can occur anywhere on the body.

When the bacteria is introduced under the skin through injury or obstructed gland, it begins to release toxins as cell damage occurs. This middle area of the abscess liquefies and grows, creating tension and inflammation under the skin. This pressure is extremely painful and most skin abscesses must be opened and drained to fully remove the infection. Sometimes, inner loculations form; these are pockets that form within the larger abscess cavity when the pus slides under a layer of tissue which then reseals creating another enclosed pocket of infection. Figure 11.1 illustrates the current treatment for skin abscesses.

*The content of Sections 11.1–11.3 and Figure 11.2 is taken from the Master's thesis of Yang (2013).

M.B. Privitera: Contextual Inquiry for Medical Device Design.
DOI: http://dx.doi.org/10.1016/B978-0-12-801852-1.00011-3

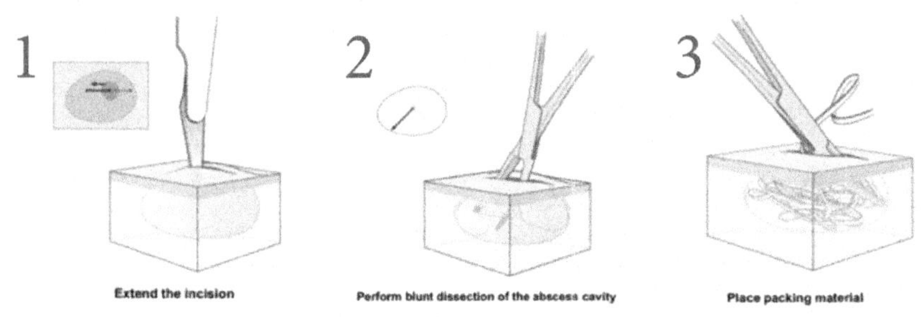

Figure 11.1 Current incision and drain procedure for the treatment of skin abscesses (Herbert et al., 2012).

The goal of this study was to capture the entire clinical procedure of abscesses, termed incision and drainage (IND), from the time of patient presentation to clinical decision-making, treatment, and subsequent gauze removal. This included a detailed description of the equipment used to prepare and complete IND treatment and the exploration of clinician perspectives, contraindications, and variances of procedure across a large clinical team.

This program built on a simple clinical annoyance brought forth by Dr. Arthur Pancioli wherein each attempt at packing an abscess using forceps and gauze tape, the tape would retract out of the abscess, therefore making no progress toward adequate packing. This method is the current treatment and is detailed below.

To enable this program, all clinical immersion activities by the design and research team were conducted at the University of Cincinnati's Medical Center (UCMC), Center for Emergency Care (CEC). All subsequent design work was completed as part of the academic Medical Device Innovation and Entrepreneurship Program (MDIEP), housed within the UC College of Engineering and Applied Science.

11.2 METHODOLOGY

In order to gain background knowledge in abscess treatment and physiology, the team conducted a thorough literature review on abscess. This included collecting scientific literature on the treatments. The team then developed models of abscess in the lab, conducting interviews and observations within the CEC, and conducting collaborative ideation sessions. Next, a comparative analysis of existing technique with novel device solutions was completed. Finally, the business opportunity was defined through access to the hospital supply chain and an analysis on current associated costs.

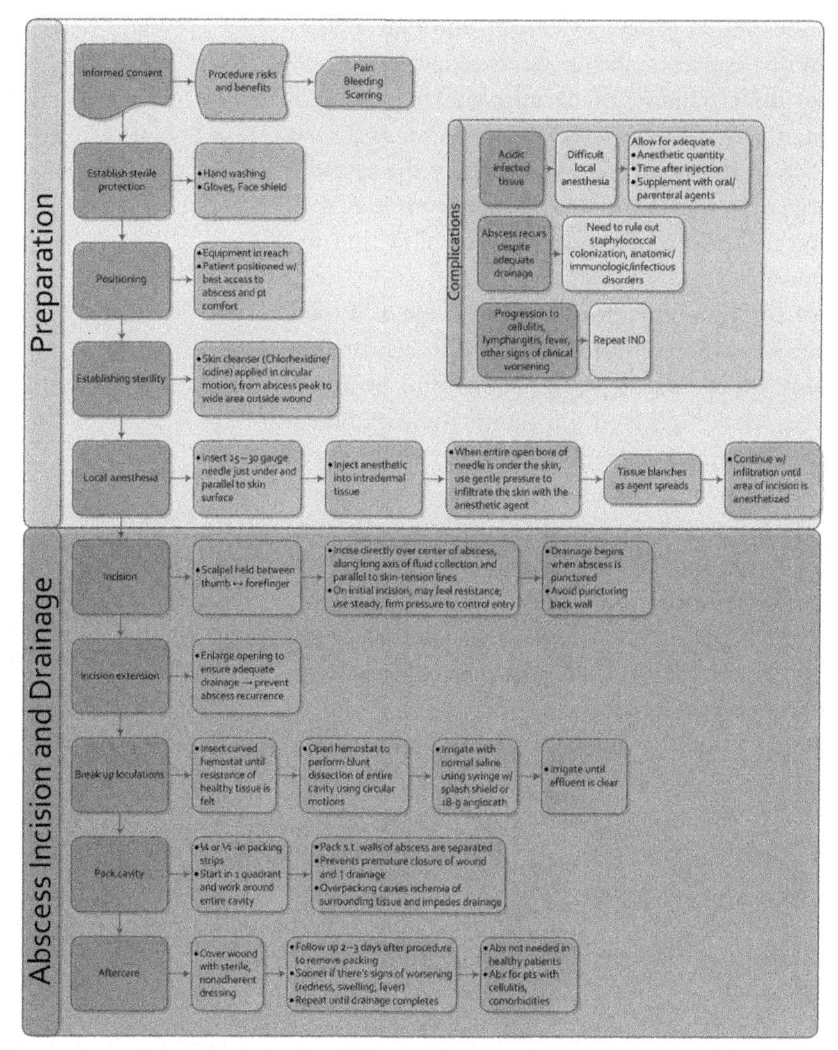

Figure 11.2 Abscess treatment according to literature.

In order to familiarize the team with abscess treatment methods and best practices according to peer-reviewed journal articles and emergency medicine texts, the following procedural flow chart was developed (Figure 11.2).

This chart formed the foundation of a formal CI study. All questions and target observations were determined from this exploration.

The collaborative team of researchers and designers developed a full study protocol and data analysis tool. The CI study recruited patients presenting the CEC, minor care department with indication for IND.

The entire clinical process from initial patient evaluation through follow-up care was observed. With consent, the patients experience was recorded. To provide further understanding of decision-making and specific challenges that consented patients who were evaluated for IND but not treated were also included in the study. The approach of the research team involved was to be placed "on call" for several days from 8 AM to 9 PM, coinciding with clinical shifts and physically stationed within the minor care unit whenever possible.

Data analysis consisted of coding events from video footage and transcribing field notes. This data was entered into a Microsoft Excel database for review and further analysis. Procedural steps, devices used or not used, clinician and patient behaviors, challenges, and mitigations were gathered. The results of the study are presented in various manners including a procedural map (Figure 11.3) and included in a formal design history file.

Concurrently, the design team was working to generate potential solutions to the observed and communicated challenges. This work was completed in close collaboration with users and utilized low-fidelity prototyping as well as rapid prototyping functional models for advanced assessment. Usability testing was completed on all substantive design iterations and the design optimized for performance.

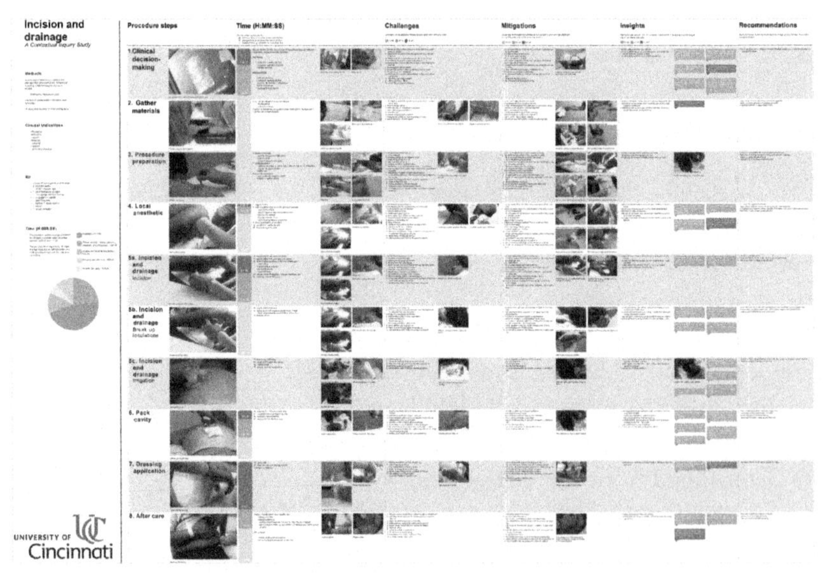

Figure 11.3 Map (Weber, 2014).

11.3 THE RESULTS

Figure 11.4 illustrates the decision-making process from patient presentation to the decision to treat an abscess with an IND procedure. This chart represents the median decision tree; dependent upon individual practitioner opinion one or more decisions may be more or less important.

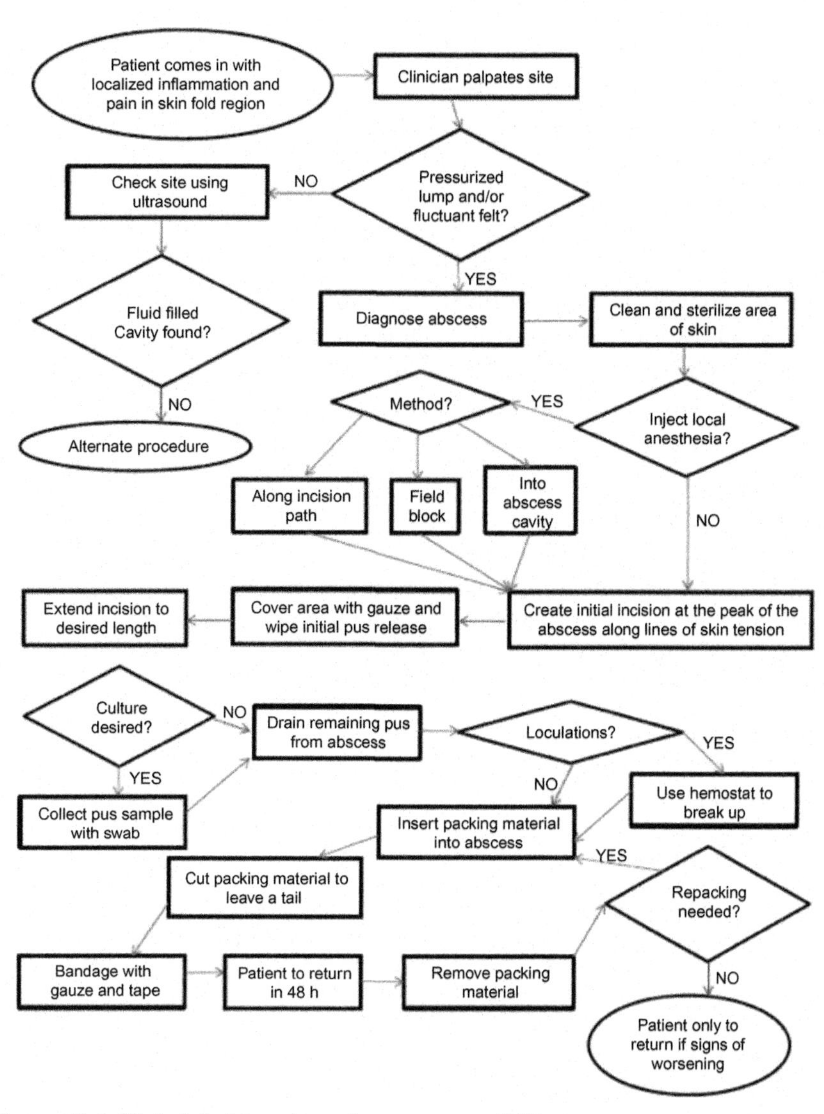

Figure 11.4 Clinical decision charts for treatment of IND.

Specific insights gained from the CI study included the following:

1. *Patient presentation*: the most common location was the chest, specifically the axilla. Extremities were also noted as common during interviews. Regions with significant skin folds (axilla, genitalia gluteal region) were frequent sites for abscess formation. Though locularity of abscess was frequently cited in literature, it was noted as difficult for clinicians to evaluate during physical examination. When completed, clinicians used ultrasound. Upon confirmation of indications for IND, only 63% of the clinicians decided to use the dedicated kit. Rather they gathered supplies from various resource locations.
2. *Tools used for IND treatment*: Gloves, Lidocaine, injection needle, sponges, flush syringe, ¼" gauze packing strips, scalpel, drape, and tape were the most commonly used tools for abscess treatment. The kits available in the CEC are made by their personnel and contain the following: ¼" gauze packing strips, 2 betadine swabs, 1 roll of 2" paper tape, 1 package 4 × 4 gauze sponges, 1 (10 cc) syringe, 1 (25 gauge) needle, 1 blunt fill needle, 1 #11 scalpel, scissors, and 1 curved hemostat. The hemostats are noted as recyclable however they are typically discarded in the sharps bin along with the needles, scissors, and scalpel.
   a. *Anesthetic injection* was viewed as most painful part of the procedure; patients were typically warned of this reality.
   b. *4x4's* are used throughout the procedure and become difficult to manage if using more than one.
   c. *Packing strips* were used in every treatment of IND. The amount of material used was inconsistent across the clinicians and based on interview of low importance. Only important to pack, not necessarily the amount of material used. Tools used to insert the packing strips are detailed below (Table 11.1).
   d. *Scalpels* were used in all IND treatments to incise the abscess. Additional uses included testing the area for sensation following anesthesia, gauging the depth of penetration based upon residual blood on the blade.
   e. *Drapes* are used to capture any sanguineous or purulent discharge as well as hold the IND instruments. Drapes were commonly tucked into patient clothing to keep them in place or used as a collection bin for waste material.
   f. *Tape* was most often used to keep the final dressing in place; however, it was also applied as a skin retractor.

Table 11.1 Comparison of Benefits and Downfalls of Current Tools Used in Abscess Packing

Tool	Benefit	Downfall
Curved Kelly forceps/ hemostat	• Grabs and inserts tape easily • Provides length for moving tape inside the wound and breaking up loculations • Common tool in EDs	• Tape can snag on the teeth which can cause the tape to retract from the wound • End is very pointed and sharp which can cause patient discomfort, and possibly further damage to wound
Toothed forceps	• Grabs and inserts the tape easily • Common tool in EDs	• Tape can snag on the teeth which can cause the tape to retract from the wound • Small forceps do not allow much length to access deep wounds or to be used to break up loculations • Hard to open and close device while in the wound; movement causes pain and discomfort to patient
Tweezers	• No teeth, so tape rarely gets snagged (minimal to no retraction of tape) • Common tool in EDs	• Hard to grip onto the tape because very small point of contact between device and tape • Small forceps do not allow much length to access deep wounds or to be used to break up loculations • Hard to open and close device while in the wound; movement causes pain and discomfort to patient

g. *Other devices*: ultrasound and otoscopes were used to determine the size of the abscess and differentiate pockets of purulence from other possibilities. Permanent markers were also used to draw a circle on the skin to indicate abscess location to indicate the spread of their infection.

3. *IND procedure steps* included gathering supplies, donning personal protective equipment, preparation, anesthesia, incision, breaking loculations, irrigation, packing, and aftercare instructions. Details of each are presented below:
   a. *Personal protective equipment* included gloves and face shields. Not all participants wore face shields citing that inconvenience of locating and wearing prevented them from doing so. Some of the physicians who started the procedure wearing a face shield removed it prior to procedural completion.
   b. *Preparation* included positioning of the patient to maximize visibility and access as well as the positioning of devices in order to maximize accessibility. Instruments were placed on chairs, on the patient, or other niches.
   c. *Anesthesia* required an averaged a total of 4 min and seven sites of injection per procedure. During interview, clinicians felt 2–4 injections would be adequate, therefore demonstrating a discrepancy between what is reflected upon and what is actually done. The average wait time for anesthesia was 7.5 min for lidocaine to take effect.
   d. *Incision* of the abscess required an average 1.5 min to complete and clinicians made 6 incisions per IND. During interview, clinicians reported the average number of incisions to be 1–2 while noting patient pain for this step to be low. In the majority of cases observed the initial incision line was extended to enable further drainage.
   e. *Breaking loculations* required an average 1.5 min to complete using a curved hemostat. The hemostat is used with multiple insertions per procedure. The identification of breaking up clinicians commented on all loculations. Additionally, the practice of squeezing the abscess was inconsistent across practitioners. Some attendings were adamant that squeezing is a potential cause for spreading infection deeper into the tissue whereas others commented that squeezing helped with visualizing the superficial border of the "pus pocket." The pressure applied, on top of the already pressurized state of some abscesses, resulted in the expulsion of sanguineous material onto the clinician.
   f. *Irrigation* was performed in 40% of the observations and mentioned in the majority of interviews. When used, there is quite a bit of irrigation to the abscess, almost pressurizing the cavity to push out any residual sanguineous fluid.

   g. *Packing* insertion required the use of hemostats to advance the packing into the cavity; however, in some instances a (Q-tip) cotton swab on a stick was used. In this case, the clinician explained the cotton swab had better grip and the incision was too small for hemostat insertion. The insertion of the packing material and determining the length was problematic in almost half of the observations.
   h. *Aftercare* instructions were provided in all cases. Patients were instructed to change their dressing and in some cases patients were instructed to remove the gauze tape after a certain period of time.
4. *Challenges and mitigations*: listed below are a sample of the challenges and mitigations that were identified in this study. This list is not comprehensive rather those key challenges which had impact on device design.
   a. Finding equipment and materials required clinicians to scavenge from other patient rooms or kits.
   b. Without an accurate measure of abscess volume, clinicians are required to use visual estimations and/or conceptual figures.
   c. Packing falls out during insertion results in the clinician using one hand to hold the packing in place while the other hand gathers more packing material to insert. Another clinician commented they make a flower-shaped bolus of packing material to start packing the abscess, thus creating a packing material anchor.
   d. Clinicians commented the use of tactile feedback to determine if all loculations were broken and to determine adequate packing. Tactile feedback was also used to determine healthy from non-healthy tissue surrounding the abscess.
5. *Design insights* included the following areas of improvement in developing a device to deliver packing material:
   a. Precut the packing material based on common sizes of abscess, perhaps by location.
   b. Kit contents may include skin retractor, dressing materials, prefilled syringe with lidocaine, 4×4 gauze, flush syringe, scalpel, and a delivery device for packing material—only materials used directly in the procedure.
   c. Packing device should be held by one hand.
   d. Critical factors of success (Table 11.2).
6. *Reimbursement analysis* was conducted in close collaboration with UCMC and the overall estimated kit costs were generated. Below is a detailed description of findings (Table 11.3).

Table 11.2 Critical Factors of Success for Abscess Packing Device

Factor of Success	Importance
Device creates consistent method of packing tape in abscess cavity.	Will reduce the risk of abscess cavity closing up and causing complications.
Device reduces packing time.	Will decrease IND procedure time that can reduce pain to patient and cost.
Device cost is competitive with current method and similar devices on the market.	Indicates that hospitals will be willing to purchase the device/method.
Device meets all safety criteria (FDA, other regulatory bodies).	Device can be commercialized.
Device is adopted and used by many surgeons in IND procedure.	The more surgeons who adopt the device/method, the larger the market and return on investment.

Table 11.3 IND Procedure Reimbursement Data from UCMC

	I&D—Simple/ Single	I&D—Complicated/ Multiple
CPT codes	CPT 10060	CPT 10061
Physician billing charge	$242	$435
Hospital billing charge	$432	$588
Cases at University Hospital during the 2010–2011 FY	758	677

Source: Gubser et al., 2015

By combining the National Epidemiology with the Reimbursement Data from UCMC, an overall market for the IND of abscesses in the emergency rooms across the nation was estimated.

Average cost of an IND procedure:

$$\begin{aligned}
&\frac{n_{\text{simple}}}{n_{\text{simple}}+n_{\text{complicated}}}(\text{Physician charge}_{\text{simple}}+\text{Hospital charge}_{\text{simple}})\\
&\quad+\frac{n_{\text{complicated}}}{n_{\text{simple}}+n_{\text{complicated}}}(\text{Physician charge}_{\text{complicated}}\\
&\quad+\text{Hospital charge}_{\text{complicated}})\\
&\quad=\frac{758}{758+677}(\$242+\$432)+\frac{677}{758+677}(\$435+\$588)\\
&\quad=\$356+\$483\\
&\quad=\sim\$839\,\text{per case}
\end{aligned}$$

By using the weighted average of cost per case (above), and multiplying it by the national epidemiology data, it can be estimated that the overall healthcare costs for abscesses alone runs at more than $2.7 billion.

Based on the cost for the current Abscess Packing Kit provided by UCMC CEC, the current kit costs the hospital $18.26 with all contents disposed of upon procedural completion (Table 11.4). This information will determine target costs of an entire kit with inclusion of a novel packing material delivery system.

7. *Comparative device analysis* was completed for the purposes of inspiration and ideation. The devices included in this analysis provided insights into other devices that absorb fluids within the practice of medicine.

   Table 11.5 demonstrates various devices currently available, a brief description of the device, the benefits each device has for use and the downfalls if the product embodiment were adopted directly for abscess treatment.

8. *Design development* and iterative design was continued throughout the research phases. The following need statement guided each team:

   "To design a device that can dispense a variable amount of the current standard packing tape, assisting in packing subcutaneous abscesses"

The following design considerations and requirements were developed based on clinical observations and physician input:

- Device will be operated by physicians
- One-handed operation is needed

Table 11.4 Current Abscess Packing Kit Content Costs (UCMC)

	Order Number	Cost
¼" gauze packing strips	ref # NON255145	$1.66
Betadine swabs	catalog # 260286	$0.04
Tub 4 × 4 gauze sponge	catalog # NON21426	$0.25
100 cc syringe with luer lock tip	ref 309304	$0.07
25 g 1 ½" needle	reorder # 305127	$0.04
Blunt fill needle	ref # 305180	$0.08
#11 scalpel	ref # 04511	$0.91
Suture scissor	reorder # 175IFG001 $1.90	$1.90
2" paper tape	None given	$0.25
Curved Kelly forceps	None given	$13.00
Total Kit Cost	$18.26	

Table 11.5 Comparative Technology Assessment

Company	Product	Description	Benefits	Downfalls
Shippert	Rhino Rocket	• Tampon-like gauze pack and inserter • Used to control nose bleeds	• Quick and easy to use • Inexpensive • Easy insertion • Bolus design so reduced procedure time and less pain for patient	• Does not use current packing material (but could substitute) • Bolus design—design physician want variable amounts of tape to be inserted • Not currently marketed and approved for wound packing
ArthroCare ENT	Rapid Rhino	• Tampon-like gauze around tube that is inflated with air to conform to shape of area • Used to control nose bleeds	• Quick and easy to use • Easy insertion • Conforms to shape of anatomy • Bolus design so reduced procedure time and less pain for patient	• Does not use current packing material • Bolus design—design physician want variable amounts of tape to be inserted • Not currently marketed and approved for wound packing • Does not seem to allow for drainage, just puts pressure on area and absorbs fluid

Source: Herbert et al., 2012; Bauldauf et al., 2014 and Gubser et al., 2015

- Variability of gauze deployment is important
  - One yard is considered enough gauze to give enough variability to the procedure
- User feedback on the amount of gauze dispensed
- Procedure with the device should be an improvement on the current procedure (i.e., faster, easier, less patient discomfort during the procedure)
- Disposable device which fits in the current abscess packing kit
- Gauze that is deployed does not retract back into the device
- General abscess incisions for drainage are ~1 cm

In conjunction with prototype design, the requirements were explored to determine the design specifications followed by an assigning of priority (need, want, nice to have) (Table 11.6).

Figure 11.5 illustrates initial product concepts developed based on the design requirements and exploration in low-fidelity prototyping. These concepts feature a button spring-loaded release to automated tape dispensing and an extended arm to break up loculations.

Table 11.6 Design Requirements to Design Specifications with Priority

Design Requirements	Design Specification	Priority Level
Device may be operated under single handed use Device must fit in current abscess packing kit	Device shall be less than 6" in length and at least 1" wide	Want
Device must contain a variable amount of gauze	Device must be able to contain at least 1 yard of standard ¼" gauze strips	Need
Device must deliver gauze out of the end effector without kinking or bunching within the device	The distance between end effector of device and drive wheel must be less than 2"	Need
Device end effector must fit in a general abscess incision (~1 cm)	Device must have an end effector no wider than 0.4"	Need
Device must have an end effector that is atraumatic to target tissue	Device end effector contains rounded surfaces with a radius of at least .03" (Comparable to Kelly forceps/hemostat blunt tip)	Need

Source: Herbert et al., 2012; Bauldauf et al., 2014 and Gubser et al., 2015

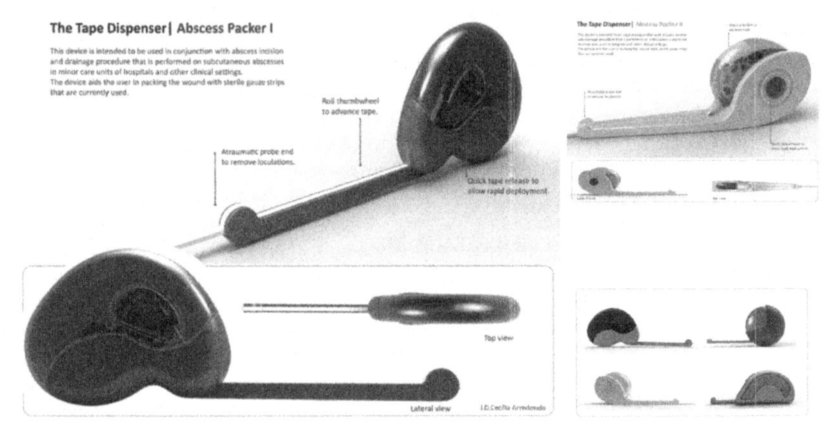

Figure 11.5 Product concepts developed by industrial design (Arredondo, 2014).

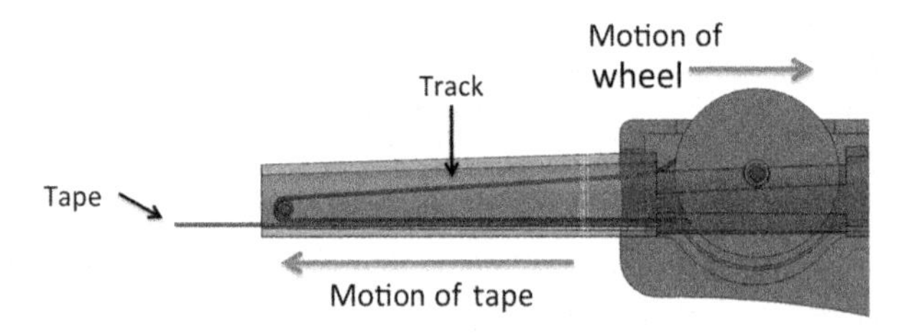

Figure 11.6 Second generation functional design (Gubser et al., 2015).

In developing a functional model of these design concepts, control of the gauze across the extended member proved problematic. As a result, the design team added a track and housing around the tape (Figure 11.6).

This design was then tested in a model simulating true abscess conditions. Typical abscess location can be located deep within muscle tissue. A full literature search was conducted to discover what others had used to simulate conditions. This search yielded few results. The design team then used several types of polymer materials (cellophane wrap, plastic baggies) and filled "pockets" with food products (yogurt, cottage cheese) to form the shape of the abscess. These pockets were surgically implanted subcutaneously into dissected chicken breasts (Figure 11.7).

A total of 18 models were produced with two abscesses each for the purposes of evaluation of the current technique and critical review of prototypes. Unfortunately, for the physician, implanting the abscess

Figure 11.7 Abscess simulation model used to demonstrate current treatment methods (Herbert et al., 2012).

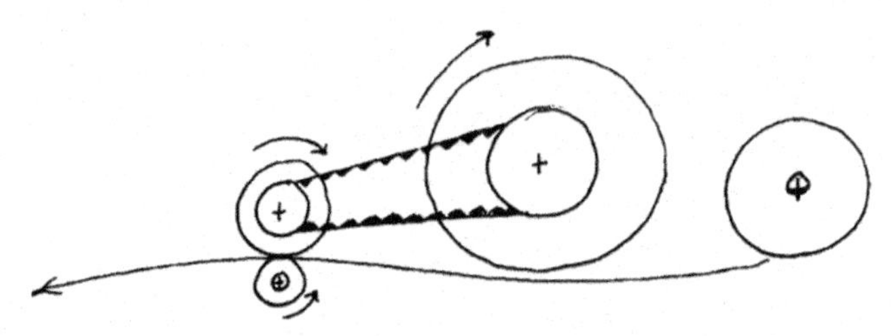

Figure 11.8 Simple pinch wheel mechanical design (Gubser et al., 2015).

model, or chicken breast augmentation, there is no reimbursement or market opportunity.

From this simulation, the amount of variance between clinicians was discovered as well as the preference to control the dispensing of the tape. Functionally, control of the tape was problematic unless delivered as a bolus as the gauze tape frequently kinked within the mechanism and housing. The mechanical design was revised to increase the surface area contact of the gauze with the internal drive mechanism (Figure 11.8).

Additionally ergonomics were explored through the assessment of other handheld tools (Figures 11.9 and 11.10). This design concept was produced using rapid prototyping techniques (Figure 11.11) and the mechanical design optimized for functionality (Figure 11.12).

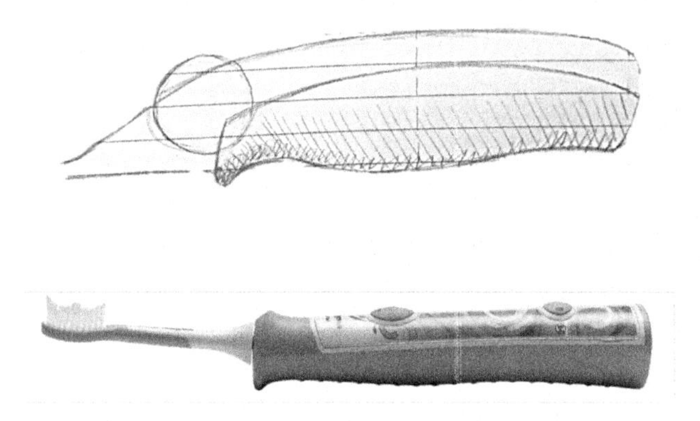

Figure 11.9 Handle design based on ambidextrous toothbrush design (Gubser et al., 2015).

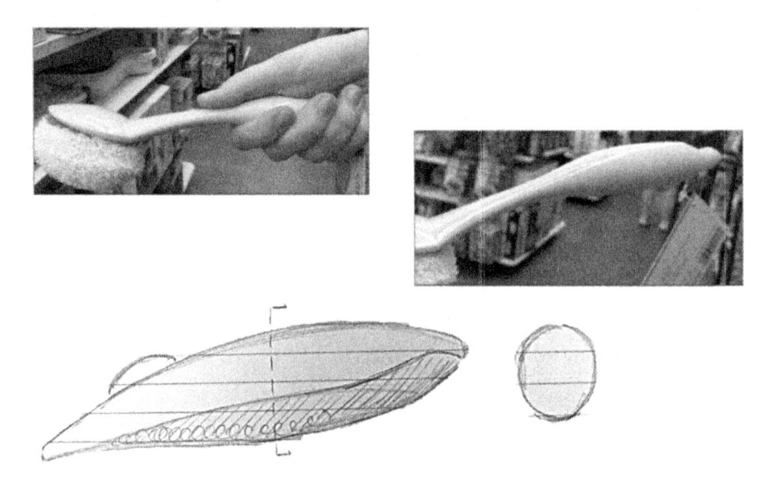

Figure 11.10 Additional handle form inspiration (Gubser et al., 2015).

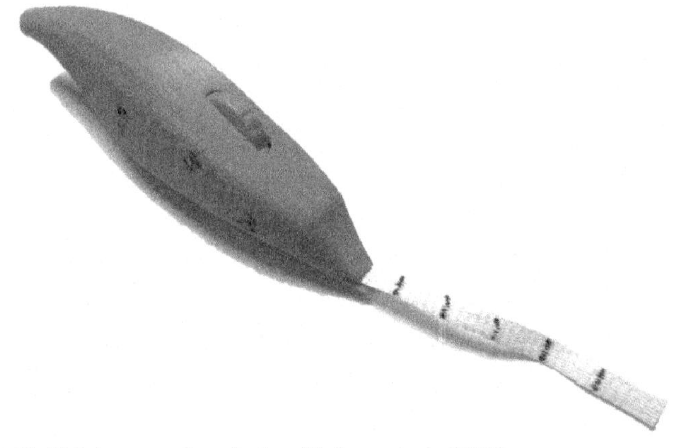

Figure 11.11 Third generation design (Gubser et al., 2015).

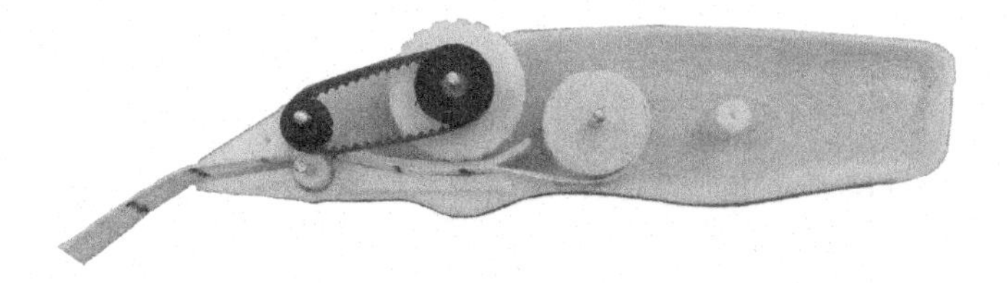

Figure 11.12 Third generation mechanical design (Gubser et al., 2015).

11.4 CONCLUSIONS

This case represents a design program that integrates a robust research phase coupled with iterative design and the production of both a "looks like" and functional models. The collaboration between designers, engineers, and clinical practitioners continued to prove invaluable as each iteration and design decision was explored.

Based on this research, there is no current device dedicated to the treatment of abscesses. There is a clear need as indicated in a study by Taira et al. (2009) in which over a 10-year period, the number of abscess cases documented at the Emergency Departments (EDs) more than doubled from 1.2 million and in 2005 just under 3% of patient visits to the adult EDs were for the treatment of abscesses. A study performed on the IND techniques in EDs found that variation was so great that there were many instances where the IND procedure was not standardized even within a single institution (Schmitz et al., 2013). Variability in the packing can lead to clinicians using inefficient methods which can be more time consuming to perform and more painful to the patient than a more streamlined process.

This successful program identified key features and cost constraints that must be considered in the device design. This includes the following:

- Ability to be used with one hand
- Ability to keep the wound open after packing
- Easy to insert and remove
- Material does not fall out of wound
- Versatile sizing/ability to dispense a variable amount of packing tape
- Overall costs of device developed in consideration of reimbursement limitations

As a result of the efforts put forth, the UC filed a US patent application in 2012 and this device is under evaluation for commercialization.

ACKNOWLEDGMENTS

This project was completed by students and faculty involved in the MDIEP at the UC. This program brings together biomedical engineering students, industrial design and business students, and faculty from UC's College of Medicine in order to solve current clinical problems both large and small. This award-winning program works in close collaboration with the medical device industry and its curriculum is directed as producing highly effective medical device development professionals.

The UC Forward Program sponsored by Provost Beverly Davenport fosters university progress through collaboration and innovation. It is through their generous funding that this work was possible. A special thanks to Dr. Arthur Pancioli who initially came forth with this clinical problem and the UC's Department of Emergency Medicine for enabling this work. Lastly, this work would not have been possible without the academic achievements of both undergraduate and graduate MDIEP students. This includes significant contribution from the following students: Samantha Weber (BSID), Adeline Yang (MS), Cecelia Arredondo (MDES), Danielle Herbert (BSBME), Taylor MacDonald (BSBME), Kaylan McClary (BSBME), Thomas Bauldauf (BSBME), Katie Hunt (BSBME), Laura Tessing (BSBME), Sophie Kinkle (BSBME), Dylan Neu (BSBME), and Cayley Gubser (BSBME).

REFERENCES

Arredondo, C., 2014. Medical Device Innovation & Entrepreneurship Program Design History File "Abscess Packer Design". Department of Biomedical Engineering and School of Design, Masters Program, University of Cincinnati.

Bauldauf, T., Hunt, K., Ingram, L., Kinkle, S., 2014. Medical Device Innovation & Entrepreneurship Program Design History File "The Pakaderm 2". Department of Biomedical Engineering, University of Cincinnati.

Gubser, C., Neu, D., 2015. Medical Device Innovation & Entrepreneurship Program Design History File "The Pakaderm 3". Department of Biomedical Engineering, University of Cincinnati.

Herbert, D., MacDonald, T., McClary, K., 2012. Medical Device Innovation & Entrepreneurship Program Design History File "The Pakaderm". Department of Biomedical Engineering, University of Cincinnati.

Schmitz, G., Goodwin, T., Singer, A., Kessler, C.S., Bruner, D., Hollynn, L., et al., 2013. The treatment of cutaneous abscesses: comparison of emergency medicine providers' practice patterns. West. J. Emerg. Med. 14 (1).

Taira, B.R., Singer, A.J., Thode, H.C., Lee, C.C., 2009. National epidemiology of cutaneous abscesses: 1996 to 2005. Am. J. Emerg. Med. 27, 289–292.

Weber, S., 2014. Medical Device Innovation & Entrepreneurship Program Design History File "Examples of Contextual Inquiry". Department of Biomedical Engineering and School of Design, University of Cincinnati.

Yang, A., 2013. Integrating ethnography, physiology, and clinical decision making to inform medical device designs. Master's thesis. Department of Physiology. University of Cincinnati, College of Medicine, Cincinnati, OH.

CHAPTER 12

Quick Reference and FAQ

Mary Beth Privitera
University of Cincinnati, and Know Why Design, LLC, Cincinnati, Ohio, USA

Contents

12.1 INTRODUCTION

This chapter serves as a quick reference guide for conducting a contextual inquiry (CI) study and answers frequently asked questions (FAQs). The entire process of conducting a CI study in medical device development is briefly presented along with key graphics to assist the team jump-start their program and to act as a means to help explain the process to those on the periphery of the research. A collection of FAQs are include covering

M.B. Privitera: Contextual Inquiry for Medical Device Design.
DOI: http://dx.doi.org/10.1016/B978-0-12-801852-1.00012-5

common topics such as the definition of CI, process of conducting a study, data analysis, developing insights for medical device design, and tools used in CI studies. Next, it answers FAQs such as the means confirming of gathering the right information, who typically conducts CI studies, variances in CI study approaches, gaining global perspectives, and overall return on investment to the business. See Chapters 1–7 for additional detail and Chapters 8–11 for specific case study examples.

12.1.1 What Is CI for Medical Device Development?

CI is a systematic study of people, tasks, procedures, and environments in their work places in order to define new product development opportunities or improve existing devices. CI, for medical device development, is used to create a body of information about the habits, devices, constraints, and/or systems delivery of care. This is especially important as those who design and engineer medical devices are not necessarily users of them.

Figure 12.1 illustrates the process of conducting a CI study. The process begins by seeking background information, setting study objectives, and planning the research. Appropriate approvals must be sought prior to data collection and the research team may need additional training and credentialing. A study protocol and discussion guide should be developed and reviewed. When prepared, the team should pilot a site-visit observation then make any necessary adjustments to study materials. While there is no ideal number of observations, typically 7–12 observations are required but the number may vary greatly; the process must include enough data to identify patterns of behavior; sample size will vary based on the complexity of the tasks, the breadth of user profiles, the variety of use environments, and the quality of access for the research. Once data has been collected it is analyzed in detail and design insights are generated.

12.1.2 How Does it Impact Device Development?

Table 12.1 highlights the how CI can be leveraged at each phase of the medical device development process. While design control systems vary between organizations, the phases presented are those typically found throughout the medical device industry and follow regulating agency requirements. Note: The impact is written in *italics* with a description of activities in the phase included.

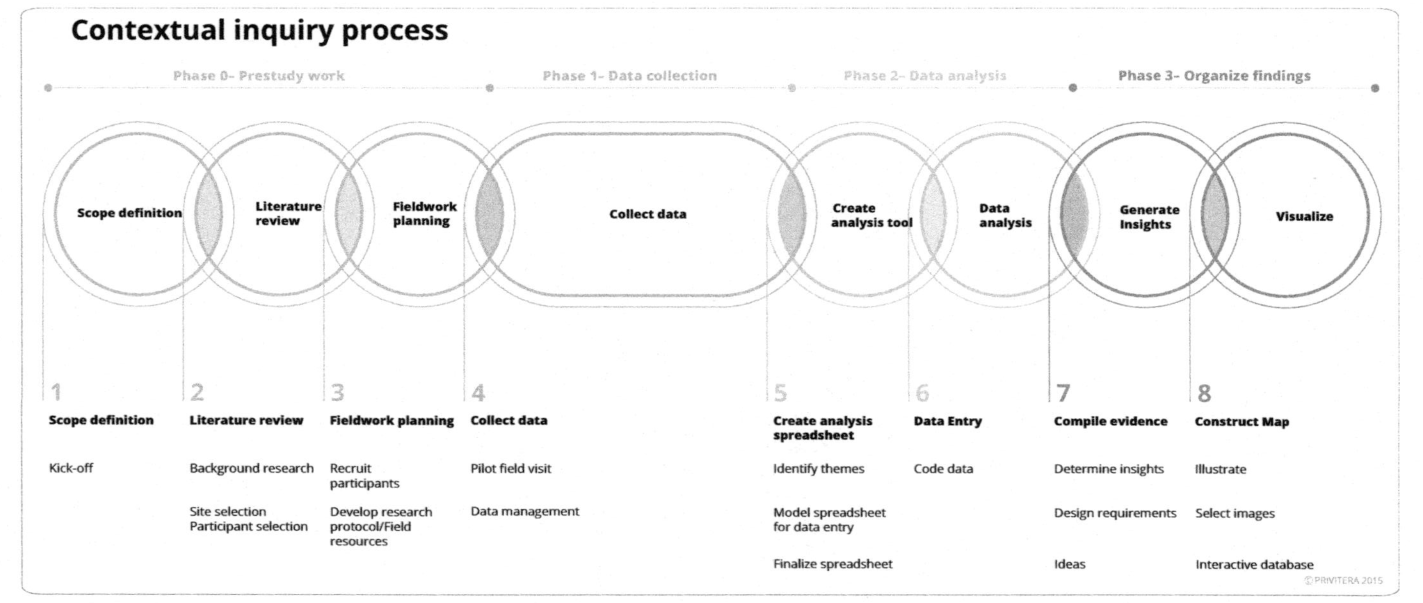

Figure 12.1 CI process for medical device development.

Table 12.1 Impact of CI Studies Defined for Each Phase of Device Development

Phase	Title	*CI Impact* Within Each Phase
0	Exploration	*Define new opportunities, define user behaviors, define use environment, define social structure around device use.*
1	User needs	*Develop fundamental design requirements in the words of the user based on their values.* The market potential, use, intended user, and use environment are fully described and the technical requirements are included. A need statement is a concise description of a goal (what the device needs to do) but does not indicate how to achieve the goal. It can also include qualitative targets and use descriptions.
2	Design input	*Inform storyboards of device, identify partners for preliminary formative evaluations, and other techniques which further design exploration and definition.* During this phase, best practices suggest that formal design verification and risk analysis begins with emphasis on preliminary testing of conceptual designs and identification of potential application risks that should be mitigated through design.
3	Detail design	*Inform design regarding desired user interface, inform risk analysis, and identify partners for further formative evaluations* as the design is finalized into dimensional concepts. Manufacturing considerations and further risk and usability assessments take hold.
4	Design output	*Provide root traceability of human factors* application at the onset of the program while conducting studies focused on assuring the design is robust. Preparing for regulatory submission and manufacturing pilots.
5	Medical device	*Improve usability* and reduce likelihood of use errors, reduce issues submitted to postmarket surveillance. One important point is the additional requirement of postmarket surveillance by regulating agencies in order to provide real-time assessment of the risks and benefits of a medical device. This requirement further emphasizes the need to focus early design efforts with the user in mind.

Taking a methodical approach to truly understanding users, their workplace, habits, and motivations can impact every stage of medical device development from ideation through postmarket surveillance.

12.1.3 How Do I Begin a CI Study?

*True wisdom is knowing what you do not know.—**Confucius***

At the onset of a study, capturing what a team already knows, or thinks they know, can provide the baseline on the clinical problem that is the focus of the study; formalizing their hypothesis is an important first step. From here, the team can determine those questions that build upon, refute, and/or confirm their hypothesis. The focus of the study can be to research and understand how an existing device is used (or not) in the field or how a new innovation may fit within existing practices. Regardless of study objective and ultimate product development goal, a thorough understanding of the existing landscape, current customer data, and any background science should be explored before drafting the CI protocol.

12.1.4 Do We Need IRB Approval?

The short answer is, maybe. Research will ideally be conducted in the actual use environment but simulations (*in vivo* or *in vitro*) may be necessary alternatives. Assuming the work is done in a hospital or other clinical care setting, the research team must understand the requirements of the specific institution; a solid draft of the study protocol will be necessary to facilitate the dialogue. Every institution has unique requirements for meeting HIPPA laws and assuring patient privacy; they further have various tolerances for access and data capture. Some may argue that any systematic observation of human behavior is human subject research and as such should undergo review; that perspective is an over generalization but some institutions take that perspective. If there is any question, get the review; it will open doors and make the study more like other studies which happen in that clinical environment.

When proposing a research protocol, it is often helpful to remind the hospital that CI studies are focused on generating new medical devices and not generalizable knowledge; the results are typically not published, but in fact, the results are typically tightly held by medical device companies.

12.1.5 What Methods Are Typically Used in a Study?

CI is based on objective observation, of users while they perform their tasks coupled with unbiased analysis.

During observations a research team's primary role is often to sit quietly watching and taking notes and recording actions. The protocol may have the team ask the user to "think aloud" during the procedure and walk through what it is that they are doing and why they are doing it, but the dialogue must not modify the actual user actions. In some situations, it may be impossible to catch a user performing a task due to the *ad hoc* nature of the clinical situation such as an emergent condition; in those cases the research team may need to develop a high-fidelity simulation to replicate the details of the actual application. Optimally, the research team would have the opportunity to actively engage in the procedure themselves under user observation in order to gain firsthand understanding.

While observing, the research team should take note of the steps in the task, the interactions of the people and devices, as well as details on the social interactions of the users. The team must be as comprehensive as reasonably possible gathering images of the specific targets as well as the use environment, recordings of comments and data, and notes on unique observations.

Planned conversations or semistructured interviews may be used in CI studies to facilitate user thought. These may include asking a user to recall particular instances of a given situation thereby providing a narrative of experience. The research team may want to use props such as emoticons or prototypes to gather opinion during the conversation regarding a device concept and/or technology. The semistructured interview should be an active interview with natural dialogue skipping from one topic to the next based on the user response. Field guides are often used to assure all questions are answered and to help assure the researcher does not begin to bias the discussion or wander outside the scope.

During the conversation the research team should strive to establish a positive rapport with the user, be prepared with the questions, and only ask one question at a time. The research team should practice medical nomenclature and be comfortable with the language. Lastly, the most important skill during the conversation is to actively listen to the user responses and reaffirm understanding without leading.

12.1.6 Do I Need Any Special Training or Certifications to Watch Clinical Care?

Yes. Each institution will provide guidance on the necessary requirements; common items include TB test results and flu shots. Many institutions now require additional training and certifications through third parties such as RepTrax and VCS. Often this training is the same training a medical device sales representative would be required to have in order to gain access to the hospital and gives the research team valuable knowledge on terms, expected conduct, and safety hazards.

12.1.7 What Do I Do with All This Data?

Proper planning will include development of databases that help organize the information into searchable files and folders. Information should be identified and generally sorted by common themes that are dominant behaviors, ideas or trends seen throughout a site visit. The specific data should be further coded into actionable categories (Figure 12.2):

- Reaction code—as observed by the team and indicated as such
- Descriptive code—inventory of tools or tool use description
- Emotional codes—opinion
- Sequential code—order of procedure
- Value and belief code—ideas and concepts that users hold as truth

From here the themes and codes can be brought together to further gain understanding and insight (Figure 12.3).

A spreadsheet or dedicated software can be used to analyze the data and sort it based on different variables. Case observations should be coded independently then brought together for comparison. From here the data can be condensed and the following actions can be undertaken:

- Draft maps of experience can be generated.
- Interactions of all user types can be identified.

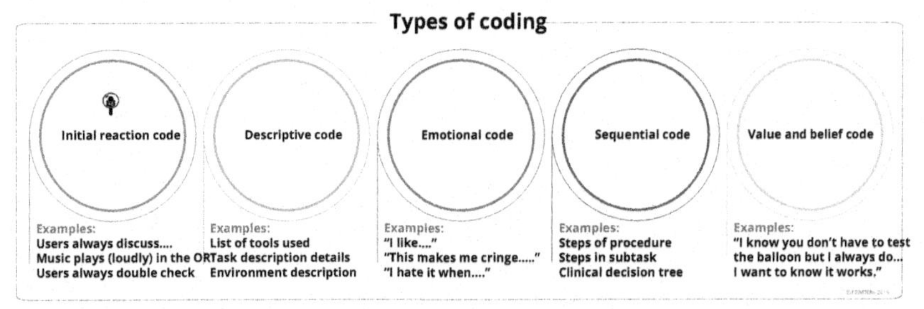

Figure 12.2 Types of coding and examples.

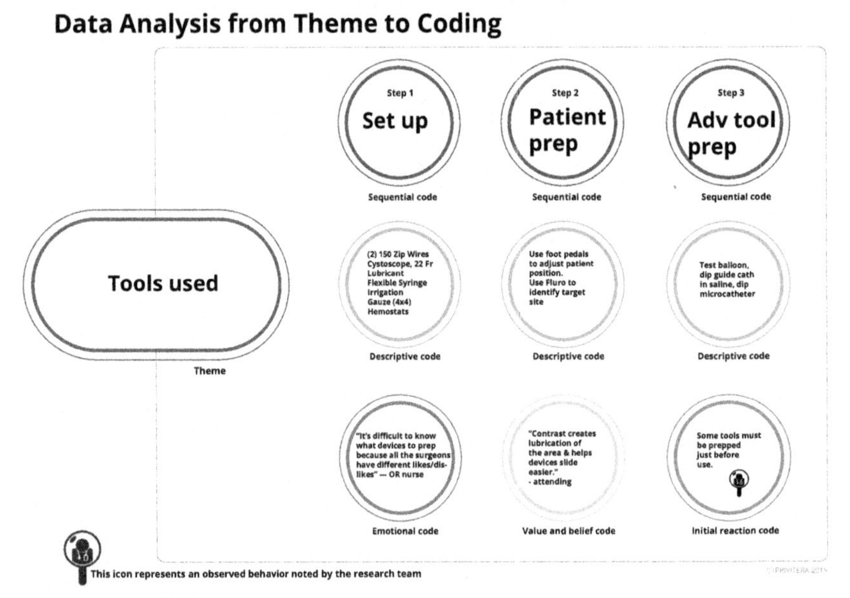

Figure 12.3 Grouping a theme together with multiple pieces of evidence that has been coded.

- Interrelationships of interaction points can be generated.
- Interrelationships of user experience can be generated.
- Specific recommendations for design can be generated.

12.1.8 How Do I Develop Insights that Are Meaningful?

The entire product development team should work together and immerse themselves in the data, seek to understand it, digest it fully, and then take time for reflection. Answer the "SO WHAT?" in order to have "AHA" moments. Keep in mind that words matter and that carefully constructed design information can have lasting impact. Common definitions are given below:

Design Information	Definition
Insight	A *design insight* is user behavior/s or quotes that are identified as important considerations in designing a new device yet may not be readily actionable.
User need	*User needs* are straightforward, can be broken down in very specific manners that relate directly to primary functionality (e.g., to drain X or to remove Y). Other needs are more complex and require a deeper understanding of

	the condition/environment/situation and relate to more subjective factors such as to diagnose, ensure success, encourage, or aid.
Design recommendations or considerations	*Design recommendations* or *considerations* are those high-level attributes that establish a general need but require more specific statements to establish a full set of measurable criteria.
Ideas or concepts	*Ideas* can include novel *concepts* that have been communicated either verbally, written in words or sketched.
Design requirements	*Design requirements* are the functional attributes that enable the team to convert ideas into design features.
Design specification	*Design specifications* are measurable device criteria, have meaning, a source as to why it is there and be associated with a test methodology.

The process of moving from insight to specification is illustrated below in Figure 12.4. It is not a discrete set of steps rather an evolutionary process with each advancement of design information and subsequent design building off one another.

A CI study will produce high-level attributes that establish a general need but require more specific statements but need further information in order to establish measureable criteria. Figure 12.5 highlights the four main areas of design consideration.

These include:

- *Social relationships*—the relationships between stakeholders, how they interact, and overall attitudes towards one another.
- *Specific actions*—the discreet tasks being performed such as opening a package, handing a device to clinical teammate, etc.
- *Motivations of device use and/or the procedure*—the data that drives clinical decision-making.
- *Overall context of use*—the environment of use.

In generating insights based on the data gathered, a development team should do as shown in Table 12.2.

12.1.9 What Are Tools Used to Communicate CI Study Results?

It depends on what information the team is communicating, for what purposes, and to whom. The collected data is rich with detail. Some of

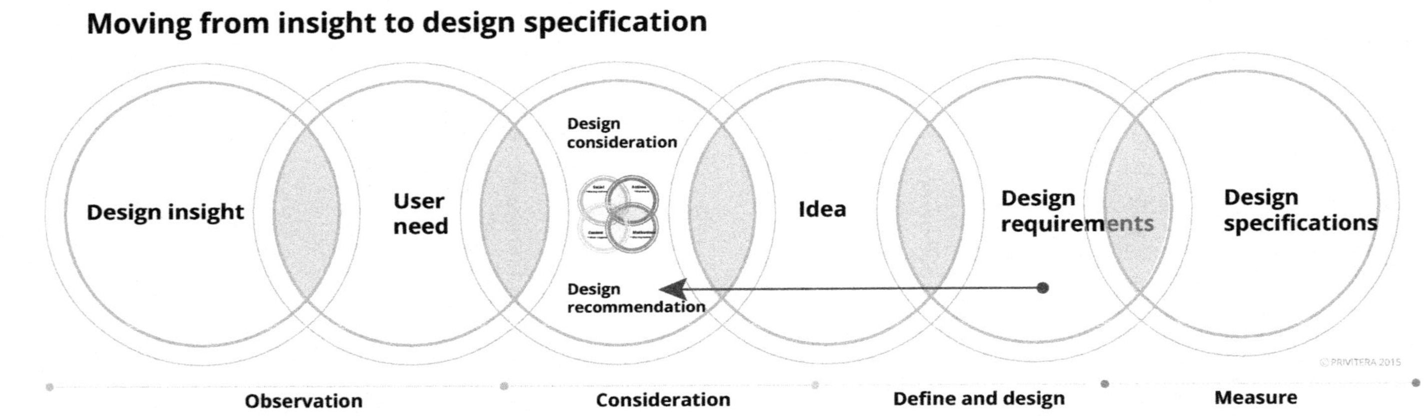

Figure 12.4 Transition from insight to specification.

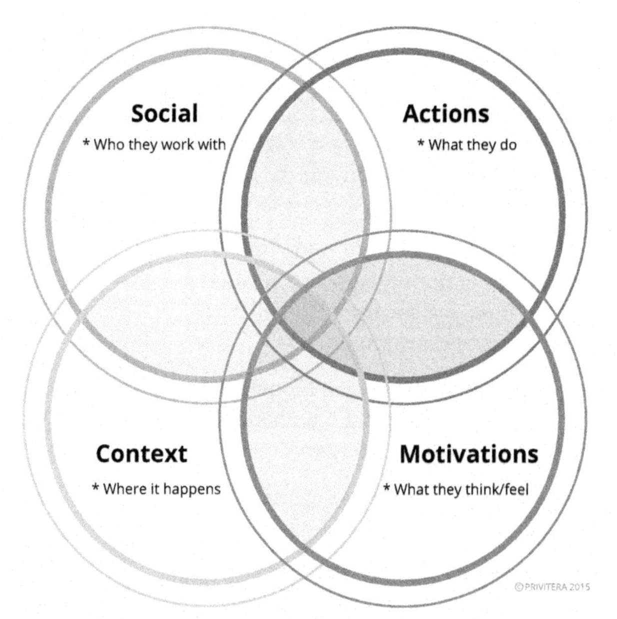

Figure 12.5 Origins of design recommendations and considerations.

Table 12.2 Prompts for Generating Themes and/or Insights

Try This:
• Review all the data generated: the themes, specific observations, challenges, and mitigations during device use. • Generate statements on the obvious behaviors or opinions demonstrated. Then look beyond the obvious. • Reframe the problem. • Look for connections between behaviors and opinions. • Look for connections between product design and use behaviors. • Look for connections between product design and clinical goal. • Find analogous technologies or situations. • Define various user types. • Review all the data generated: the themes, specific observations, challenges, and mitigations during device use.

this detail needs to be exploited whereas in other instances detail must be removed. Table 12.3 presents typical uses.

Clinical providers, responsible for the training of others, may use the final results of in their pedagogy as a tool to enhance training or courses. Product development teams may use the final procedure maps to (1) uncovering new opportunities for device development and as a

Table 12.3 Types of Communication with Purpose Description

Type of Communication	Purpose
Images/ photographs	Communicates a specific point in time, easy to highlight individual elements such as user position and/or disproving false assumptions.
Illustrations	Best used to communicate complexity. By removing external elements found in a photograph or the ability to draw cross-sections of device/anatomy, detailed device tissue interactions can be explored.
Videos	Videos provide rich and comprehensive data. Significant events can be edited for retrieval and then embedded in presentations of interactive database references.
Storyboards	Simple illustrations or images of procedure steps or patient/user journey.
Spider diagrams	Visual display of quantifiable data such as time for each step in a procedure.
Interactive database	Digital database wherein each step within a procedure can be explored in depth with images, illustrations, and video.
Procedure maps	Full two-dimensional visualization of a CI study (Figure 12.6). Typically in poster form and displayed in product development area.

reference for development; (2) communicate the procedure with other members within the medical device organization; (3) encourage deeper conversations with clinical providers in the science behind the device use; and (4) assess marketing claims that relate to surgeon use or patient outcomes.

12.1.10 How Do I Know I Have the Right Information?

If you are learning, you have the right information. Information gathering should be focused on objective learning and can impact every aspect of the device design and development. It should advance the team's knowledge of the clinical science, the personalities and culture of the users as well as provide key information on the context of use. This should be seen as a means for involving the entire product development

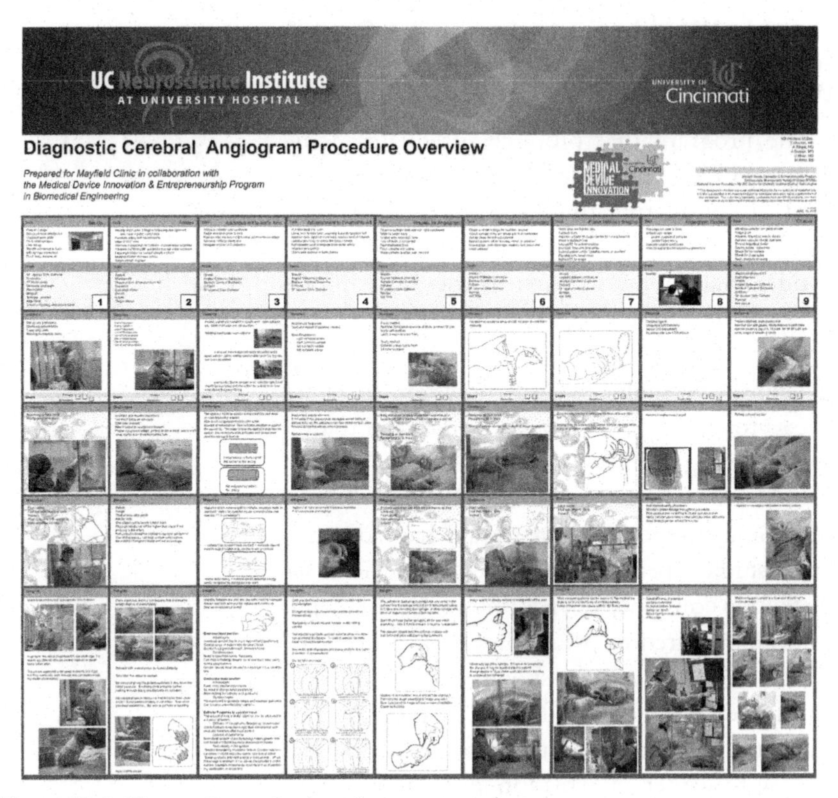

Figure 12.6 Diagnostic cerebral angiogram procedure map.

team and creating a common understanding that will facilitate effective team communication and motivation.

12.1.11 Do I Have to Do a Formal CI Study? Can't I Just Go Watch Without All the Data Hassle?

The formality of the study depends on the time and budget constraints (if completed externally to the organization). Simply watching is helpful but often product development teams make assumptions about what they see and why it exists, which can create or reinforce bias. Without a systemic means of gathering design-related information, making objective logical decisions with strong evidence becomes challenging. The use of a systemic CI approach, tailored to the need, should be seen as a cost-effective learning process that yields data that becomes a tool supporting the development process as a long-term reference for postmarket surveillance and product validation.

12.1.12 What Kind of People Would I Go to Conduct a Study? Who Is Trained to Do This? What Are Their Credentials? What Do I Look for?

There is no formal academic program that produces graduates dedicated to exploring the front end of medical device development. The types of disciplines involved in conducting these studies include: biomedical engineering, industrial design, experimental psychology, mechanical engineering, and social scientists. Each of these disciplines has their strengths. Ideally, a multidisciplinary team conducts the study.

There are no formal credentials to conduct a study however; there are specific requirements for access to clinical care. These include registration and training with vendor credentialing agencies.

The research team should be familiar with AAMI TIR51:2014 Human Factors Engineering-Guidance for CI and should have demonstrated experience conducting studies specifically for medical devices. CI studies are routinely completed for consumer devices; however, the depth in analysis of science may not be required for those products. An organization seeking a supplier of CI research services should seek consultants with clear capabilities in scientific research. In addition, they should be able to demonstrate the translation of their findings into relevant design information. It is relatively easy to gather information and summarize it. Interpreting and translating it for the purposes of design requires talent and expertise.

12.1.13 Is the Process Any Different Depending on the Setting such as a Home Healthcare Device Versus In-Hospital Use Device?

No. The overall process is the same however the recruiting efforts will be different. Home health studies may go through marketing research recruiters using screeners more readily than those studies completed in the hospital for specialized procedures. Additionally, home health site visits may take longer in the actual site visits in the field, as study participants may be particularly conversational.

12.1.14 How Does a CI Study Consider Global Perspectives?

The purpose of a CI study is to go ask rather than assume. This includes gaining a global perspective. Each culture around the globe has unique qualities that may impact the design, marketing, and reimbursement of a device. As such, gaining a global perspective is warranted, particularly when the development team knows that cultural and economic variables play a role in the selection and use of products. Abbreviated studies in international sites are means for cost-effectively supplementing domestic observation providing the team with firsthand knowledge of those variables that need to be mitigated.

12.1.15 How Do I Show that it Adds Value in Return on Investment to the Business?

The success of any medical device is dependent on the decisions made throughout the development program. The device must work as intended, be safe, usable, and the overall cost must be appropriate for the market. Each of these must be balanced during the development process. Beginning the process with a CI study with involvement of the device development team brings shared vision and responsibility. A properly conducted CI study should yield a product that is more intuitive, requiring less in-service/sales support, and accelerating the adoption curve. A product that properly mitigates procedural challenges should have fewer clinical complaints and have lower long-term burden on the quality of the organization. A clearer set of product specifications at the onset of the development program should facilitate a more efficient development cycle and reduce the overall development costs. And most importantly, the proper definition of the overall product goals should increase the success rate of the program and help assure the project capitalizes on the market opportunity while also helping cull concepts early that are misdirected.

INDEX

Note: Page numbers followed by "*f*" and "*t*" refer to figures and tables, respectively.

J

K

L

M

N

O

V

W